WHO CARES ABOUT PARTICLE PHYSICS?

T0201860

WHO CARES ABOUT PARTICLE PHYSICS?

Making Sense of the Higgs Boson, the Large
Hadron Collider and CERN

Pauline Gagnon

Indiana University,
USA

OXFORD
UNIVERSITY PRESS

OXFORD
UNIVERSITY PRESS

Great Clarendon Street, Oxford, OX2 6DP,
United Kingdom

Oxford University Press is a department of the University of Oxford.
It furthers the University's objective of excellence in research, scholarship,
and education by publishing worldwide. Oxford is a registered trade mark of
Oxford University Press in the UK and in certain other countries

© Pauline Gagnon 2016

First Edition published in 2016

Published in the United States of America by Oxford University Press
198 Madison Avenue, New York, NY 10016, United States of America

British Library Cataloguing in Publication Data

Data available

Library of Congress Control Number: 2015959281

ISBN 978–0–19–882627–9

Printed and bound by
CPI Group (UK) Ltd, Croydon, CR0 4YY

To my parents, Colette Perron and Paul Gagnon, who gave me so much, and to my friends Kate Hieke and Cath Noyes, both gone way too early.

Acknowledgments

If you are like me, you may like reading the acknowledgments at the beginning of a book just to glimpse at what the author might have gone through. This being my first book, I was especially afraid of feeling isolated after relocating 500 km away from CERN, where I had spent the previous 19 years. However, thanks to the incredible support I received from colleagues and friends, it was quite the opposite. Even when I was sitting alone at my desk, they were by my side, albeit virtually and electronically, revising one or several chapters, or offering me advice and encouragement. This was so much so that, in the end, I had the impression of having achieved some teamwork. It was really reassuring for a person like me, used to working with 3000 other people on the ATLAS experiment. I was not completely surprised, though, knowing that the vast majority of my colleagues are just as eager as me to share the chance we have to participate in the unique adventure taking place at CERN right now.

I want to thank profusely (and in alphabetical order) Sylvie Brunet, Natalie Garde, Penny Kasper, Narci Lorenzo and Pascal Pralavorio for reviewing the whole book or most of it, offering me invaluable advice and suggestions about both the content and its presentation. Being able to count on their assistance made a whole amount of difference. Several other colleagues and friends also checked one or several chapters for scientific accuracy or clarity. I am therefore very grateful to Alexandre Arbey, Sudeshna Banerjee, Thomas Cocolios, Michael Doser, Monica Dunford, Louis Fayard, Jules Gascon, James Gillies, Geneviève Guinot, Vincenzo Iacoliello, Marumi Kado, Clara Kulich, Nazila Mahmoudi, Sophie Malavoy, Judita Mamuzic, Giampiero Mancinelli, Django Manglunki, Markus Nordberg, Marie-Claude Pugin, Yves Lagacé, Pierre Savard and Andrée Robichaud-Véronneau for their time and for their excellent suggestions for improvement. Their generous help and support were thoroughly heartwarming. Thanks to them, there should be many fewer mistakes, and the text is much more fluid. They took time off from their evenings, their weekends and even their holidays to assist me. I am immensely grateful to all of them.

Huge thanks also go to Kate Kahle for believing in this project from the start and for her sustained support up to the end. I am also very grateful to all my friends who joined me for e-lunches on Skype. I also wish to thank Jean-Marc Gagnon from MultiMondes, my editor for the French edition, for his extremely enthusiastic response when I first contacted him as well as Ania Wronski from Oxford University Press for her professionalism, her sound advice and her patience throughout the editorial process.

Special thanks go to my mother for her Larousse dictionary and her Bescherelle grammar book, and for passing on to me her taste for well-done work. Finally, I wish to thank my partner, Marion Hamm, for her patience, her love and her multiple encouragements, especially for insisting I stray away from my computer to get some fresh air. Otherwise, I would certainly look like a Higgs boson in the middle of winter.

Chapter Summaries

Chapter 1: What Is Matter Made Of?

What are the smallest grains of matter, and how do they interact to form all the matter that we observe around us? The Standard Model is the current theoretical model that describes all these particles and their interactions, giving us a clear representation of the material world. It can even predict the behavior of these particles to a very high level of accuracy. Each one of these particles also comes with its own antiparticle. However, nearly all the antimatter has mysteriously disappeared from our Universe.

Chapter 2: What about the Higgs Boson?

The media have relayed the message that the Higgs boson gives mass to fundamental particles. In fact, three elements are needed to generate the masses of fundamental particles: a mechanism, a field and a boson. The Brout–Englert–Higgs mechanism is a mathematical formalism that describes by means of equations a real physical entity, the Brout–Englert–Higgs field. This field is simply one of the properties of our Universe, like space and time are. The Higgs boson is an excitation of this field, just as a wave is an excitation of the surface of the ocean. Finding the Higgs boson proved the existence of this field.

Chapter 3: Accelerators and Detectors: The Essential Tools

Producing Higgs bosons was one goal of the Large Hadron Collider, or LHC, at CERN. It first accelerates protons to almost the speed of light, and then brings them into collision. This accelerator can concentrate a huge quantity of energy into an extremely small point in space. The energy generated materializes in the form of particles. Most of these particles are unstable and break apart moments after appearing. Detectors located at the collision points act as huge cameras to catch the fragments of these ephemeral particles.

Chapter 4: The Discovery of the Higgs Boson

By sorting billions of events collected with the detectors operating at the LHC, physicists can extract the few events bearing the characteristics of the Higgs boson. Advanced statistical methods make it possible to disentangle the Higgs boson from all other types of events, allowing scientists to extract the needle from millions of haystacks.

Chapter 5: The Dark Side of the Universe

The Standard Model works incredibly well but only applies to a mere 5% of the content of the Universe. In turns out that 27% of our Universe is made of a strange type of matter, something absolutely mysterious called dark matter. The remaining 68% of the Universe constitutes a form of energy as enigmatic as it is unknown. But evidence of the existence of dark matter abounds. It plays an essential role in cosmology, acting as a catalyst for the formation of galaxies. We can detect its presence through its gravitational effects and by means of gravitational lenses. Several experiments are currently in progress deep underground, aboard the International Space Station and at the Large Hadron Collider at CERN hoping to catch dark matter particles for the first time.

Chapter 6: Calling SUSY to the Rescue

The Standard Model, in spite of its amazing success, has several flaws: for example, it does not explain gravity, nor dark matter. Obviously, there ought to be an even more encompassing theory, which would be based on the Standard Model but would be more far-reaching. A plausible and tantalizing theory called supersymmetry, or SUSY, is very popular. SUSY has everything to please us. It is based on the Standard Model, unifies the grains of matter with the force carriers and comes with a new particle that would be an ideal candidate for explaining dark matter. Its biggest problem is that it remains undiscovered. So is there still a chance that this hypothesis could be right? Oh, yes!

Chapter 7: What Does Basic Research Put on Our Plates?

All this research has a cost. Is it worth it? My answer is yes, without hesitation. Thanks to basic research, all humanity has a better knowledge of the material world surrounding us. This is already a lot, but there is much more if we take into account all the other benefits. Scientific activities yield a highly trained workforce. These people contribute to the development of society in many respects. The economic and technological spinoffs from engaging in basic research make it one of the best investments even in the short term.

Chapter 8: A Unique Management Model

Thousands of researchers working in cooperation without immediate supervision, free to decide when, where and how they want to work. Is this realistic? This is indeed how the big collaborations in particle physics operate. This management model favors creativity, personal initiative and the empowerment of all those involved. It relies simply on the interest shared by the entire community in bringing their experiment to a successful completion. This model could also benefit many companies.

Chapter 9: Diversity in Physics

More women now undertake a career in physics than was the case a few decades ago. However, at CERN only 17.5% of the scientists are female, although the situation is continually improving. Why is this so, and how could the situation be made better? Women are not the only underrepresented group in this field. Science has everything to gain by being more inclusive in terms of gender, race, sexual orientation, religion and physical ability. Diversity in science leads to greater creativity.

Chapter 10: What Could the Next Big Discoveries Be?

As a conclusion, I pull out my crystal ball to predict what can be expected in terms of discoveries in the coming years. In particular, the restart of the Large Hadron Collider at CERN in 2015 at higher energy opened the door to new discoveries. These breakthroughs could revolutionize our conception of the material world surrounding us.

Contents

Introduction

Many of you may have heard about the Higgs boson and the Large Hadron Collider (LHC) at CERN, the European laboratory for particle physics. But who really has a good grasp of these concepts? Here is a book written with you in mind, using terms as simple as possible and intended for every interested person who is not a specialist. It aims at explaining clearly what the Higgs boson is, as well as other key topics in particle physics, to make it accessible to as many people as possible. Hence, every interested reader should be able to discover the fascinating world of particle physics without letting mathematics or excessively detailed explanations obscure the subject. A good dose of curiosity should be enough, and no prior knowledge of advanced mathematics or of scientific concepts beyond high school level is required.

This book neglects several historical and mathematical details to retain the essential points, or at least I hope it has done the latter. We physicists often tend to obscure the subject with our obsession with being absolutely right. In contrast, although this book is scientifically correct, it is aimed instead at being accessible. There are neither equations nor complex calculations. Any person armed with a minimum of interest can thus go through it without stumbling too much, and share the passion that motivates thousands of scientists engaged in this type of research.

However, not wanting to short-change the most inquisitive readers, I have supplied all the information those people are entitled to expect. These details have been gathered together beside the main text in boxes, to lighten the reading for those mostly interested in the global picture. Several more specialized books are available on the market that will allow the eager reader to go into greater depth.

If you are a curious person who simply wants to know how part of your tax money is being used to subsidize research in science, this book is for you. You too can benefit from the extraordinary knowledge that ensues from research in particle physics. And if the pleasure of understanding better the world that surrounds us is not enough in your opinion to justify the enormous sums invested in research, a whole chapter explores the impressive economic and social impact of basic research.

If a paragraph seems too difficult, keep on with it or just move on to the next section. The level of complexity does not increase over the course of the book. Each chapter can, essentially, be read independently of the others. So if you get a little lost, which might occur occasionally, rest assured: all chapters end with a brief summary where the main take-home message is recapitulated. These summaries will allow you to skip a section or even a whole chapter to suit your taste. I hope that in this way everyone will find what he or she needs, be they curious retirees seeking out new knowledge, students eager to discover the world, or my friends and family and their neighbors.

The book begins with an explanation of the goal of particle physics and a description of the world of fundamental particles. We then get to the heart of the subject to discover the nature of and unique role played by the Higgs boson. We will see how fundamental particles are produced at the LHC and how physicists detect them. Then we will take a giant step, going from the infinitely small to the infinitely large to realize that all the current knowledge in particle physics explains only 5% of the content of the Universe. Everything else remains to be discovered. This suggests that another, much broader and more encompassing theory could soon replace the current model.

Source: © Particle Zoo (with permission).

The success of particle physics experiments rests on a unique management style, where the management team for each experiment coordinates activities following preagreed mechanisms rather than by imposing views and directives on people. Hence, thousands of scientists coming from tens of different countries work with a high level of autonomy, without directives or immediate supervision, united simply by a common objective: discovering how the material world works. Diversity means creativity, although particle physics still has much to gain by welcoming more people of different gender, sexual orientation, race, religion and physical abilities.

The book ends with the near future and explores what could be the next big discoveries in particle physics in the coming 10 or 20 years. We are most likely on the verge of a huge scientific revolution. I hope that my book will save you from running the risk of being left behind. You too can discover what the Higgs boson is about and other essential topics to better understand some of the key issues in physics today.

1

What Is Matter Made Of?

What are the smallest constituents of matter that exist, and how do they hold together to form all the matter that we see around us? Answering this question is precisely the purpose of particle physics. This branch of physics aims at finding what are the smallest existing grains of matter, those that cannot be broken into smaller parts, and how they interact with each other.

Imagine a place where everything material was built from Lego bricks (see Figure 1.1). Then, if I were to ask you the question "what are the smallest parts of matter here?" the answer would be quite simple. It would be enough to break up various objects to see what were the smallest bricks coming out of them. From that, one could deduce what the fundamental particles were in a world made completely of Lego bricks. Everything could be built from these elementary bricks. The same goes for all matter: it is all made of "elementary bricks," except that these smallest, indivisible pieces are too small to be seen. Moreover, it is nearly impossible to break matter up into its smallest constituents.

This question concerning the smallest grains of matter is nothing new. Numerous people have asked the same question throughout history. Two thousand five hundred years ago, Leucippus and his follower Democritus, two Greek philosophers, had the right idea when they proposed the theory of *atomism*, a doctrine stating that all matter consists of atoms and empty space. In ancient Greek, *atomos* means "indivisible," that is, unbreakable into smaller parts. Unfortunately, nineteenth-century scientists concluded too hastily that they had found these indivisible elements. The name was thus applied wrongly to what we nowadays call *atoms*. However, we now know that these atoms are composite objects that can be divided into smaller components.

The smallest grains of matter

So, what are the fundamental particles in the real world? With real matter, it is more difficult than with a hypothetical world made of Lego

Figure 1.1 If all matter were built from Lego bricks, here is how the fundamental particles would look. But in real life, it is much harder to see what are the building blocks of matter.

Source: Pauline Gagnon

bricks since we cannot easily see its smallest components. But in physics labs, physicists can. Matter is indeed made of atoms, but atoms are not fundamental. They are composite objects, as can be seen in the diagram in Figure 1.2. They have an atomic nucleus containing protons and neutrons, with the electrons forming a cloud around it. Atoms are therefore mostly constituted of empty space. To get a sense of the size of an atom, imagine the nucleus is the size of your body. The electrons would then be smaller than a hair and orbiting around you at a distance of about 12 miles (20 km). Matter is therefore mostly made of vacuum,

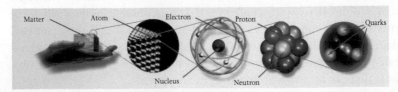

Figure 1.2 All matter is made of atoms. The neutrons and protons of the atomic nucleus are made of quarks. Inside atoms, only the quarks and the electrons are genuine fundamental particles, that is, indivisible particles. They cannot be broken into smaller components.

Source: CERN.

emptiness and some fundamental particles. We will now examine how this holds together and why matter appears to us as if it is solid.

Hence, atoms consist of other particles. Even protons and neutrons are not indivisible: they are made of quarks and gluons, the latter acting to bind the quarks together. In the end, the only indivisible particles at the heart of matter are thus the quarks and the electrons. We will come back to the gluons shortly.

The recipe for protons and neutrons

Protons and neutrons are formed from quarks. We obtain a proton by combining two *up* quarks with one *down* quark. *Up* quarks have an electric charge of $+\frac{2}{3}$, that is, two-thirds of the unit charge of an electron. *Down* quarks have a charge of $-\frac{1}{3}$. For a proton, we therefore have *up* + *up* + *down*, or $+\frac{2}{3}+\frac{2}{3}-\frac{1}{3}$, and thus an electric charge of +1. A neutron contains two *down* quarks and one *up* quark; hence its charge is $+\frac{2}{3}-\frac{1}{3}-\frac{1}{3}=0$. It is electrically neutral.

This is illustrated in the diagram in Figure 1.3 using little creatures from the Particle Zoo. Julie Peasley, seamstress by training and zoo-keeper by passion, started the Particle Zoo[1] after attending a public lecture on particle physics. I will use her particles throughout the book.

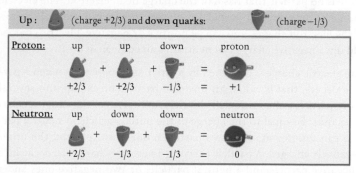

Figure 1.3 Protons and neutrons can be built from *up* quarks and *down* quarks.
Source: Pauline Gagnon and © Particle Zoo.

---

[1] All her fuzzy little particles are available for purchase on the Particle Zoo website, http://www.particlezoo.net

Electric charge

One of the best-known properties of fundamental particles is their electric charge, since it manifests itself not only at the subatomic level but also on our macroscopic scale. The electric charge of the electron is −1, and this value constitutes the basic unit of charge. The electron charge gives rise to electricity. An electric current is simply a movement of electrons within a conductor.

Electricity is very similar to the flow of water in a creek. The moving electrons are just like water drops. Each one carries a unit of electric charge. The total quantity of water passing per second gives the flow. Similarly, the total number of electrons passing per second gives the *current*. Its strength is measured in units of *amperes*, or coulombs per second. In these units, the electron charge is a mere 1.6×10^{-19} coulomb, or 0.00000000000000000016 coulomb. Hence six billion billion electrons must pass per second to give one ampere. The *voltage*, or *potential difference*, corresponds to the change in elevation: the more slope a creek has, the more energy the water has.

Scientists believed for a very long time that the electric charge of the electron was the smallest unit of charge. However, quarks have fractional values of this charge, namely exactly one-third or two-thirds of the electron charge. Why? We do not know, just as we do not know why there is no particle that has half the charge of an electron. The electric charge of fundamental particles is always specified in multiple values of the electron charge, and can be positive or negative. Electric charges add up: a negative charge can neutralize an equivalent positive charge.

The electric charge is governed by a strict *conservation* rule: when a particle decays, that is, when an unstable particle breaks up into several other particles, the sum of all the electric charges of the daughter particles must be equal to the charge of the initial particle. A neutral particle can disintegrate into two particles, one positively and the other negatively charged. A particle carrying a negative charge can decay into a negative particle and a neutral particle, or two negative ones and a positive one. Electric charge can never disappear, nor can it appear from nowhere.

The atoms

Protons, neutrons and electrons are enough to form all of the possible atoms that constitute the 118 chemical elements of the periodic table (Figure 1.4). In turn, the 118 chemical elements can be combined in diverse proportions to form various *molecules*, which are aggregates of atoms. Atoms and molecules constitute the entire visible matter that we observe around us, be it on Earth or in stars and galaxies.

In atoms, the electrons spin around the atomic nucleus. What keeps them there? It works pretty much as for a stone attached to a string that someone is spinning around.[2] The string keeps the stone in a circular orbit. If the person were to let go of the string, the stone would keep going in a straight line. As long as we hold it, though, the string exerts a force on the stone, constantly bringing it back toward our hand, forcing it to go in a circle.

Electrons are also maintained in a circular orbit around the atomic nucleus by an "invisible string." This string is nothing but the attractive

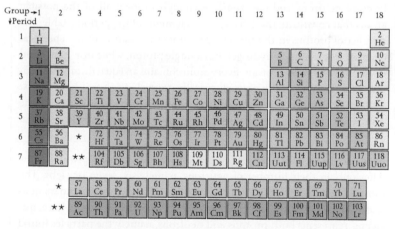

Figure 1.4 One can obtain the 118 chemical elements by mixing protons and neutrons, both built from *up* and *down* quarks, as well as electrons, in various proportions.

Source: Wikipedia.

---

[2] This analogy only holds up to a certain point: the "length" of the string in an atom is *quantized*, that is, it can only take specific values.

force between the negative charge of the electron and the positive charge of the protons in the nucleus. This force acts on the electron exactly like the string on the stone. The same goes for the planets orbiting around the Sun. In this case, though, the gravitational force is the invisible string. The matter contained in the Sun generates a gravitational force that provides the force necessary to keep the planets orbiting. All forces are like invisible strings acting upon fundamental particles or large objects. We will come back to this shortly.

To summarize: *up* and *down* quarks combine to form protons and neutrons. They, in turn, are regrouped within the atomic nucleus. By adding electrons, one gets atoms. By varying the number of protons in the nucleus, one can build the 118 different chemical elements of the periodic table. Finally, by combining atoms in various proportions, one can build all of the matter we see around us. Thus, everything we see can be built from a basic construction set containing electrons and *up* and *down* quarks.

Atoms and isotopes

Protons, neutrons and electrons are enough to form all of the possible atoms that constitute the 118 chemical elements of the periodic table: the number of protons in the nucleus determines the nature of the chemical element. For example, hydrogen has a single proton, while iron has 26 and uranium contains 92 protons. Every atom contains an equal number of protons and electrons and is thus electrically neutral. An atom that has lost some of its electrons becomes positive and is called an *ion*. By changing the number of neutrons, one obtains the various *isotopes* of one single chemical element.

For example, each of the three isotopes of carbon contains six protons. They differ only in their number of neutrons, namely six, seven or eight. The most stable and most abundant carbon isotope has six protons and six neutrons. We denote it as ^{12}C to indicate that it contains 12 *nucleons*, where 'nucleons' represents both protons and neutrons, namely the particles found inside the nucleus. The carbon isotope containing eight neutrons is called carbon-14, or ^{14}C, and is radioactive. This simply means that its atomic nucleus is unstable and will eventually break apart into smaller, more stable atoms, at a specific rate.

Atoms, isotopes and molecules (*continued*)
....

Carbon-14 is used to date plants and animals in archaeology. It is produced when cosmic rays strike nitrogen atoms in the air. A living body always ingests a well-established mixture of ^{12}C and ^{14}C. But as soon as it dies, the quantity of carbon-14 contained in the body decreases steadily, since it is radioactive and its stock is no longer replenished. Given that we know that it takes roughly 5730 years for half the carbon-14 atoms to decay, a sample can be dated simply by estimating the quantity of carbon-14 that it still contains.

The Standard Model

Over the past fifty years, scientists have developed a very precise theoretical model to describe the basic components of matter and the forces that act upon them. This model has helped us to classify all particles observed until now according to their properties. The model took shape through a close collaboration between experiments and theory. Discoveries made in physics laboratories served as the basis on which theorists could develop a logical and coherent representation of the material world. The experimental observations made it possible to confirm or eliminate various theories. Likewise, the theoretical hypotheses that were formulated guided the experimentalists in their searches.

The current theoretical model of particle physics is called the Standard Model. It rests on two rather simple ideas, which are in a way its basic principles:

First principle: all matter is made of fundamental particles.
Second principle: all these particles interact with each other by exchanging other particles.

After approximately a century of research in the field, we now know that there are 12 basic grains of matter in nature, all fundamental particles (Figure 1.5). They come in two categories: the leptons and the quarks.

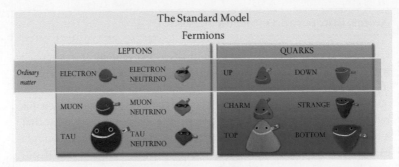

Figure 1.5 The 12 fundamental grains of matter: six leptons and six quarks. *Source*: Pauline Gagnon and © Particle Zoo.

Leptons

The best known of the six leptons is the electron. Two other charged particles, the muon and the tau, are very similar to the electron but much heavier. All these leptons carry an electric charge of −1.

The electron, the muon and the tau are each "associated" with a neutrino, namely the electron neutrino, the muon neutrino and the tau neutrino (we refer to these as the three *flavors* of neutrinos). These six particles form the lepton family. This family contains three *generations* of leptons. Each generation contains one charged lepton and its associated neutrino. This association of particles within a generation reveals itself through the formation of pairs. When an electron is produced, it always comes either with an electron neutrino or with an antielectron, a particle called the positron.

Similarly for muons and taus: each one is always produced with its own neutrino or with its antiparticle. We never observe the production of a tau together with an electron neutrino, for example. The electron and its neutrino both have a property, the *electron flavor*, and this property is subject to a conservation rule just like electric charge. When a pair is formed, one particle carries a flavor charge, and the other the opposite flavor charge. If neutrinos were absolutely massless, this flavor charge would be perfectly conserved just like the electric charge. But this is not always true, as we will see in the next section.

Just like neutrons, neutrinos are electrically neutral particles but much smaller, hence their name, which means "little neutron." Having no electric charge, they interact extremely rarely with matter. As an example, every second, 70 thousand billion electron neutrinos emitted by the Sun strike every square centimeter of the Earth's surface. Of all these neutrinos, only a handful will interact with an atom of matter on Earth. The others will pass through the Earth without even stopping to say hello!

Quarks

Besides leptons, there are *quarks*, which form the second family of fundamental particles. They come in six flavors, or types: we have already been acquainted with the *up* and *down* quarks, which are found in protons and neutrons. Then come the *charm* and *strange* quarks, and finally the *top* and *bottom* quarks. These are also called *truth* and *beauty*, although "*top*" and "*bottom*" are more commonly used. These names were chosen partly for fun, but also because the scientists who discovered them did not understand why there were so many and what distinguished them. The third quark to be discovered was the *strange* quark. It owes its name to its surprisingly long lifetime.

Nobody knows why quarks and leptons come in three generations, each one having such different masses, nor why only the first generation is needed to form atoms and, consequently, all of the ordinary matter found around us. Imagine if a Lego set contained bricks with sizes so poorly matched. Moreover, what if some of the pieces were of no use? These are some of the many still unresolved questions that particle physicists are trying to elucidate.

None of the particles of the second and third generations are found in nature, except for muons in cosmic rays. Although all these particles existed just after the Big Bang, the Universe has now cooled too much to have enough energy to produce them.[3] We can, however, generate all of them in the laboratory, and this is how we know they exist.

The mass of neutrinos

Scientists believed for a very long time that neutrinos had no mass. But this changed with the observation of a very peculiar phenomenon called *neutrino oscillations*, a process by which a neutrino of a given type, say an electron neutrino, changes into another type, either a muon or a tau neutrino.

continued

---

[3] As we will see in Chapter 2, there is an equivalence between mass and energy, which means that particles can be produced when enough energy is available.

The mass of neutrinos (*continued*)

This metamorphosis can only occur if neutrinos have a mass. Its observation has thus established that neutrinos do have a mass. Consequently, the lepton flavor charge that I have just mentioned is not perfectly conserved. But since the masses of neutrinos are extremely small, this violation is also rare. We only observe it when neutrinos travel over large distances.

Ray Davis, one of the pioneers in the study of neutrinos, was the first person to detect the neutrinos emitted by the Sun. Using a large radiochemical detector placed at the bottom of a mine in Minnesota, he established beyond any doubt that only a third of all the neutrinos emitted by the Sun reached the Earth. The number of neutrinos coming from the Sun was estimated using a theoretical model describing how the Sun produces its energy. Ray Davis dedicated 30 years of research to this measurement. His work was rewarded with a share of the Nobel Prize in Physics in 2002.

His measurement left a huge question unanswered: what was happening to the other neutrinos? The answer came from the Sudbury Neutrino Observatory (SNO), an experiment located deep in a mine in Sudbury in northern Ontario in Canada. The SNO scientists demonstrated that, in fact, some of the electron neutrinos emitted by the Sun turned into muon or tau neutrinos during their transit. This phenomenon of oscillations explains the apparent disappearance of solar neutrinos. Ray Davis's detector was only sensitive to electron neutrinos, the unique type of neutrinos produced by the Sun. But the SNO detector used heavy water, a substance sensitive to all three types of neutrino. Oscillations had already been observed in Japan in "atmospheric" neutrinos, those produced when cosmic rays strike particles of the atmosphere.

The SNO detector was able to verify that the total number of neutrinos from the three different flavors was indeed equal to the expected number of neutrinos produced by the Sun. SNO established that neutrino oscillations were taking place for solar neutrinos, confirming that they had a mass. However, their mass is so small that we have still not managed to measure it accurately, even though we know that it is different from zero. Nothing is sneakier than neutrinos! The Nobel Prize in Physics 2015 was awarded jointly to the Japanese Takaaki Kajita and the Canadian Arthur B. McDonald for the discovery of neutrino oscillations.

Nevertheless, this observation challenges the Standard Model, even though the model does not predict the mass of any fundamental particle. But

---

The mass of neutrinos (*continued*)
....

adding masses to the neutrinos in the model can be tricky, since neutrinos are so special. They are the only grains of matter with no electric charge. So what kind of particle is a neutrino? An electron is not identical to its antiparticle, the positron, because one has a negative charge and the other has a positive charge. But a neutrino is neutral. Therefore a neutrino could be its own antiparticle. This would be the only such fermion in the Standard Model. What does this mean? If a neutrino is its own antiparticle, it could be that it gets its mass differently from the other particles. Moreover, the fact that the mass of a neutrino is so small suggests that the neutrino mass is special. Hence, this raises several unanswered questions.

As we will see in Chapter 6, this is one of the many clues we have telling us that the Standard Model is flawed and that a new model must be developed. Neutrino physics is a whole branch of particle physics, and entire books have been dedicated to it. This topic will not be covered here, but interested readers could consult, for example, *Neutrino Hunters* by Ray Jayawardhana.

The force carriers

Remember the second basic principle of the Standard Model? The fundamental particles interact with each other by exchanging other particles, which are the carriers of the *forces*, those "invisible strings" I mentioned earlier. The force carriers have some characteristics that make them belong to a large class of particles called *bosons*, whereas the grains of matter, the quarks and leptons, belong to another class called *fermions* (see the box "Fermions and bosons" later in this chapter).

By exchanging these force carriers, other particles feel the effect of the force that the force carriers are associated with. This is a bit like two people skating on ice and moving in parallel. If one skater throws a heavy snowball at the other, the latter skater will feel the impact, and that will make her deviate from her initial trajectory. Likewise, the first skater will recoil from his throw and will deviate too from his initial direction. You can test this yourself: put on a pair of roller skates or ice skates and throw a heavy object ahead of you. If your skates have eliminated all resistance and friction, and if you manage not to break your neck, you will feel the recoil. The same effect is felt if you try to catch a heavy object thrown at you. This recoil is the same as what you experience when firing a gun.

The fundamental forces

There are four fundamental forces: the strong force, the electromagnetic force, the weak force and the gravitational force. The strong force is the most powerful of all but it only acts at very short distances, and only on quarks. This is what distinguishes the quarks from the leptons. The force carrier (Figure 1.6) for the strong force is the *gluon*, a particle without mass that "glues" quarks together, as its name suggests. This force is powerful enough to keep quarks inside protons and neutrons, and to overcome the electric repulsion between protons (see below). Its effects barely extend beyond the radius of neutrons and protons, however, just far enough to keep them inside the atomic nucleus. Its reach is limited to the size of an atomic nucleus, that is, 10^{-15} m or 0.000000000000001 meter.[4]

The second strongest force in decreasing order of magnitude is the electromagnetic force, carried by photons. Two electrically charged particles "feel" the presence of one another by exchanging photons.

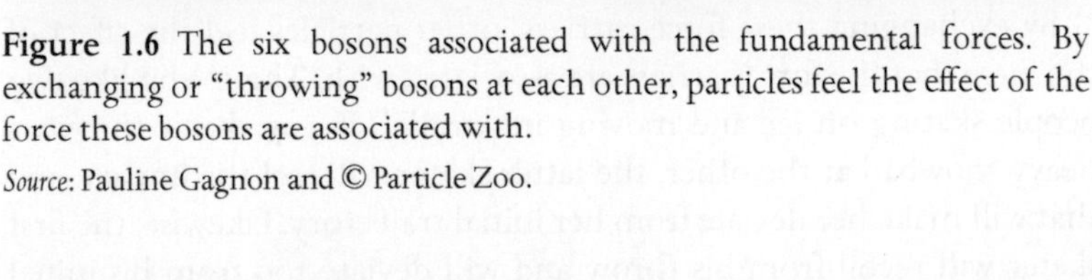

BOSONS

GLUONS	PHOTONS	W and Z BOSONS	GRAVITONS
Strong interaction	*Electromagnetism*	*Weak interaction*	*Gravitation*

Figure 1.6 The six bosons associated with the fundamental forces. By exchanging or "throwing" bosons at each other, particles feel the effect of the force these bosons are associated with.

Source: Pauline Gagnon and © Particle Zoo.

---

[4] I use the *scientific notation*, to simplify the text. For example, for 10^5 years, one takes 1 and adds five zeros to it, which yields 100,000 years. Negative exponents represent fractions. So, to transform 10^{-5} second, one starts with 1.0 and then moves the decimal point by five positions to the left. Hence, one gets 0.00001 second or 10^{-5} second, which is simpler than one hundredth of one thousandth of a second.

This is how the attraction or repulsion is established between two electric charges, depending on whether they have opposite or like signs.

The electromagnetic force, which affects only electrically charged particles, plays an essential role in all our lives. There is an electric repulsion between the atoms of your chair legs and those of the ground on which it stands. Without it, your chair would go through the floor. Atoms are mainly empty volumes, but the repulsive force coming from the electric fields generated by the electrons near the surface makes everything seem completely solid. One can visualize the effect of this electric field by imagining that atoms are surrounded by springs. To put two atoms very near each other, one would need to compress these springs. The resistance would grow bigger and bigger the more they were compressed, making it impossible for two atoms to get too close to each other. In the end, this ensures that matter, which is a collection of atoms, seems solid, compact and impenetrable, when in fact it is mostly empty as we saw earlier.

The third force is the weak force, the one responsible for the decay of particles and for radioactivity. It is carried by three bosons. The W^+ and W^- bosons both carry a unit charge. One is positive, and the other negative. There is also the Z^0 boson, which is electrically neutral. The weak force acts on all particles, both leptons and quarks. It is the only force that acts on neutrinos, if we neglect the gravitational force, given that their mass is so tiny.

The last force, the gravitational force, is the one that allows you to read at the moment while comfortably seated or lying down. Except, of course, if you are reading this aboard the International Space Station,[5] where one is in a state of weightlessness.[6] The gravitational force nevertheless remains the most mysterious of all the interactions. At the scale of quarks, it is 10^{41} times weaker than the electromagnetic force, that is, 100,000,000,000,000,000,000,000,000,000,000,000,000,000 times smaller.

---

[5] If that has ever been the case, let me know!

[6] Weightlessness occurs in this case because the International Space Station (ISS) is always falling toward the Earth. Just as in the example given earlier in this chapter of a stone attached to a string spinning in a circular orbit, the ISS would keep going in a straight trajectory if it were not for the gravitational pull of the Earth constantly bringing it back toward the Earth. This pull is equivalent to the station falling freely toward the Earth, causing weightlessness.

It is so weak compared with the other forces that its effects are negligible on the scale of particles. Literally astronomical quantities of matter are needed to feel its effects. To visualize the difference in strength between the gravitational force and the electromagnetic force, it is enough to think that a simple fridge magnet can overcome the gravitational attraction of the whole Earth: one can easily stick a small object equipped with a magnet onto a refrigerator to prevent it from falling. This fundamental property of matter is responsible for the worldwide prosperity of the fridge magnet industry.

The gravitational force is the only force without a known force carrier. Now that the LIGO Collaboration announced the first direct detection of gravitational waves (see http://paulinegagnon3.wix.com/boson-inwinter#! A-faint-ripple-shakes-the-World/c1q8z/56bcc6630cf2b4e0b62599f6) in February 2016, there is a chance these waves also have a force carrier called the graviton, which could be found at the Large Hadron Collider at CERN. Already, this discovery provides a new tool to explore the first instants of the Universe since nothing impedes the passage of these waves. The echo of the Big Bang is imprinted in gravitational waves that are still roaming the Universe to this day.

Fermions and bosons

The grains of matter, the leptons and quarks, all belong to one class of particles called *fermions*, whereas the force carriers are part of another class called *bosons*. These classes of particles are named after two famous physicists: the Italian Enrico Fermi and the Indian Satyendra Nath Bose, who studied these classes of particles. This classification is much more than a simple question of names. It denotes completely different behaviors for these two classes of particles. The two sets of particles in fact have different values of *spin*. The spin is just one more property that defines fundamental particles, like their electric charge and their mass. The spin represents their *angular momentum*, a measure related to their rotation as the name suggests.

In the world of the infinitely small, everything becomes "quantized," that is to say, that certain properties such as the electric charge, the spin or the *color* of quarks (the property that makes them respond to the strong force) can take only specific values, for example 1 or ⅓, or even ½. Only multiple values of these basic numbers, called *quanta* (hence the term "quantum physics"), are permitted. The allowed values are like the steps in a staircase: we can

Fermions and bosons (*continued*)

....

stand on the first or the second step, but not between them. If each step was 20 cm high, your height could only be given as a multiple of 20 cm.

A quantum number represents the discrete values (as opposed to continuous values) that certain quantities can take. The grains of matter, the fermions, have a total spin value of ½. This gives them two possible orientations: pointing upward, +½ or downward, −½. The force carriers, the bosons, have integer values (whole values) such as 0, 1 or 2. The fermion and boson groups obey different statistical laws. One mandatory rule for fermions is that two identical particles cannot exist in the same place in the same quantum state, that is, a state where all their quantum numbers are identical.

Electrons belong to the fermion group. If we want to put two electrons in the same place, for example in the same orbit inside an atom, one of their quantum numbers must differ. As we just saw, there are only two possible spin orientations, pointing upward or downward, that is, +½ or −½. This implies that at most two electrons can coexist in the same atomic orbit, given these two spin orientations are the only possibilities. Atoms thus have several atomic layers to accommodate all their electrons. The consequences are enormous, because all chemical reactions are governed by this organization of electrons into different orbits. This is referred to as the *Pauli exclusion principle*.

On the other hand, there is no limitation imposed on the number of bosons allowed to be in the same state. This property explains the phenomenon of *superconductivity*. What is this? Superconductivity is a state in which an electric current can flow freely, all resistance having totally disappeared. If one injects a current into a superconductor, this current circulates indefinitely. So, if your electric lawnmower were made of superconducting material, you would only need to plug it into an electric outlet once to get a current to flow in it. It would keep going indefinitely even after it was disconnected. However, although the current in a superconducting electric lawnmower could flow forever, the whole unit would still lose energy in cutting grass and through friction, eventually bringing it to a stop. Why don't we use this wonderful property more to save energy? The problem is that to become superconducting, a material needs to be cooled down to between −240 and −460 °F (−150 and −273 °C), depending on the type of material. That is hardly practical. But this is probably a blessing. Otherwise, life in summer would be hell, with lawnmowers running incessantly!

continued

Fermions and bosons (*continued*)

....

When regrouped into pairs, two electrons become a boson, since two spins of one half add up to a spin of 1 if the two spins are aligned in the same orientation, and to 0 otherwise. In a superconductor, all electron pairs are allowed to be identical. Every pair can have exactly the same quantum numbers as the other ones, as this is allowed for bosons. One can thus swap two pairs freely.

In a superconductor, all electron pairs can change position with another pair without generating any friction, and hence without any electrical resistance. All this looks very much like couples moving on a dance floor. If everybody moves in the same direction during a waltz, the couples do not collide.

But let's return to our bosons: why should the grains of matter have values of spin of one half, and the force carriers integer values? We do not know. This intriguing difference could be resolved by *supersymmetry*, as we shall see in Chapter 6.

How about antimatter?

Every grain of matter has its counterpart in antimatter. For example, the antiparticle of the electron is the positron. A positron has exactly the same mass as an electron but all its quantum numbers (electric charge, spin, electron flavor) are inverted. In this way, even particles that are electrically neutral have their antiparticles. For example, an antineutron is made of one *antiup* quark and two *antidown* quarks. Neutrinos could be their own antiparticles, although this is not settled yet. When a particle meets its antiparticle, the two annihilate, leaving only their equivalent in energy. The same goes for each one of the six quarks and six leptons: they all have their antiparticles.

In laboratory experiments, we always produce matter and antimatter in nearly exactly the same quantities, as if the two were on an equal footing. However, everywhere we look around us in the Universe, there is nearly no trace of antimatter. When some is found, it is always in minute quantities, as in cosmic rays for example. In the laboratory, one observes that there is a very slight preference for matter. But this difference is too small to explain why, for practical purposes, we only find matter in the Universe.

If matter and antimatter are produced in nearly equal amounts from pure energy, the same rule also had to apply a few moments after the Big Bang, when phenomenal quantities of energy were available. This energy materialized in the form of pairs of particles and their antiparticles, which could combine with other particles or annihilate. When and how did the matter get the upper hand over the antimatter? What happened to all the antimatter? Cosmologists are convinced that it could not be hiding in some corner of the Universe without revealing its presence. Sooner or later, matter and antimatter would meet, producing fireworks of energy that would eventually be detected. It is thus a huge mystery in physics, and many physicists are working on this puzzle. Solving this is the main objective of the LHCb experiment, which operates at the Large Hadron Collider (LHC) at CERN. Several experiments are also in progress at the antimatter factory at CERN (see the box "Antimatter experiments at CERN").

Antimatter experiments at CERN

If matter and antimatter are always produced in nearly equal quantities, what happened to all the antimatter that must have been in the Universe after the Big Bang? To attempt to answer this question, one must first verify that antimatter and matter have the same properties. CERN is supporting a vast antimatter research program with a dedicated accelerator, the Antiproton Decelerator, or AD. This is CERN antihydrogen factory. Its purpose is to compare the behavior of antihydrogen atoms (Figure 1.7) with that of hydrogen atoms. Hydrogen was chosen because it is the simplest of all atoms, having a single electron orbiting around a nucleus containing a single proton. Antihydrogen atoms are replicas of hydrogen atoms, with a positron—the antielectron—and an antiproton replacing the electron and the proton of hydrogen atoms. It would be impossible to make anything more complex in the laboratory. Even a deuterium antiatom, with one antiproton, one antineutron and one positron, is produced one million times less easily than an antihydrogen atom. Each additional proton or neutron cuts the production odds by another factor of one million.

continued

Antimatter experiments at CERN (*continued*)

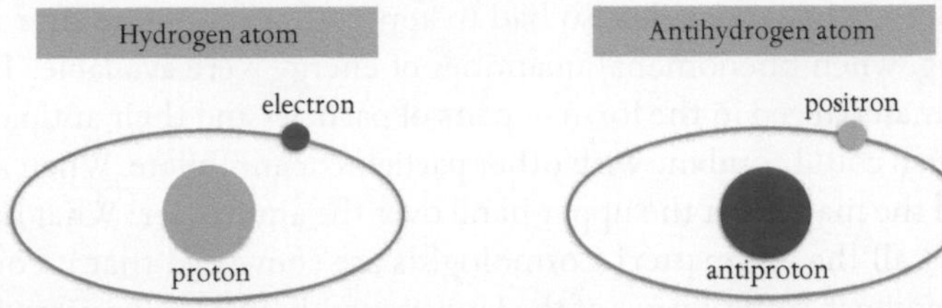

Figure 1.7 Schematic representation of hydrogen and antihydrogen atoms. *Source*: Pauline Gagnon.

All matter emits light when excited, for example when a piece of metal is heated up. The emitted light reveals the identity of the atoms producing it. A hydrogen atom emits or absorbs light of a specific frequency (or color) when its electron jumps from one orbit to another. *Spectroscopy* consists in analyzing all the colors emitted by an atom and establishing its spectrum, as can be done with a prism. Two experiments at CERN, named ALPHA and ATRAP, are concerned with the spectroscopy of antihydrogen. One can also study the "hyperfine structure" of antihydrogen. This corresponds to magnetic interactions between the nucleus and the electron spin. ALPHA and a third experiment, ASACUSA (Figure 1.8), will check this hyperfine structure of antihydrogen. In both cases, the researchers will observe the specific frequencies that antihydrogen atoms can absorb and then compare them with the known spectrum of hydrogen.

To produce antihydrogen atoms, one must first slow down antiprotons to allow them to pass slowly enough near positrons to be able to attract them and form antiatoms. The researchers combine the antiprotons with positrons using a magnetic field. This field prevents the antiprotons and positrons from coming into contact with matter, which would cause their immediate annihilation, preventing the formation of antihydrogen atoms. The final step is to move the antihydrogen atoms away from this magnetic field to be able to study their hyperfine structure. Otherwise, the strong magnetic field would decrease the precision that could be achieved. Both ALPHA and then ATRAP succeeded in trapping antihydrogen atoms in 2010, taking a giant step toward future spectroscopy studies.

Antimatter experiments at CERN (*continued*)
....

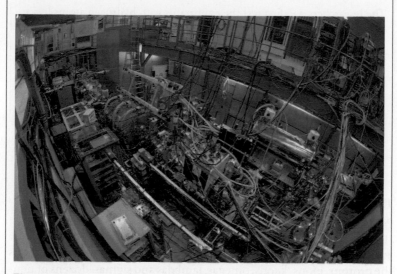

Figure 1.8 A view of the ASACUSA experiment in the Antiproton Decelerator hall at CERN. The goal of ASACUSA is to measure the hyperfine structure of antihydrogen atoms, and then compare it with that of hydrogen. The purpose is to check if matter and antimatter have identical properties to elucidate the disappearance of antimatter from the Universe.
Source: CERN.

Unlike the protons in the LHC beams, antihydrogen atoms are neutral and cannot be controlled by electric fields. However, an antihydrogen atom behaves like a tiny magnet. One can manipulate these microscopic magnets using a nonuniform magnetic field and obtain a beam of antihydrogen atoms. ASACUSA has already managed to produce such a beam. The last but not least step is to measure its hyperfine structure.

Two other experiments, AEgIS and GBAR, aim at remeasuring the gravitational constant but with antimatter. To do so, one must check whether antihydrogen atoms react like hydrogen atoms to the gravitational attraction of the Earth. Unfortunately, this cannot be accomplished simply by dropping antihydrogen atoms from the top of the Leaning Tower of Pisa, as Galileo once did with two balls of different masses for his original measurement of the gravitational acceleration constant, g. The device needed is slightly more complex.

continued

Antimatter experiments at CERN (*continued*)
....

The researchers will first test their apparatus with hydrogen atoms just to make sure everything works as expected. Once the method is well established, they will repeat the experiment with antihydrogen atoms. This will certainly be tricky but, if successful, these measurements could provide some answers. We shall know then if antihydrogen is the mirror image of hydrogen or not. This would reveal whether antimatter differs from matter and give some clues about why it disappeared from the Universe.

Particles to suit every taste

The Standard Model has greatly simplified the world of particles compared with the way it appeared in the first half of the twentieth century. The idea of quarks came long after the observation of a whole zoo of particles (more than 200 today). This includes particles such as the protons, pions, kaons, omegas, lambdas and sigmas: dozens of neutral, positive and negative particles. But in 1964, Murray Gell-Mann and George Zweig developed the quark model and greatly simplified the picture.

It was not until then that scientists realized that all these particles were made of quarks. Today, we classify all particles made of quarks in the *hadron* family (from the ancient Greek *hadros*, meaning "strong") (Figure 1.9). Some hadrons, such as pions and kaons, are formed from one quark and one antiquark. This branch of the hadron family corresponds to the *mesons* (from *mesos*, "intermediary") (Figure 1.10). Others, such as protons and neutrons, are made of three quarks; we call them *baryons* (from *barus*, "heavy").

Despite its success, however, this simple model is starting to burst at the seams. Since 2003, several experiments have detected particles containing four quarks. These *tetraquarks*[7] do not fit within the current quark model. Several theorists are hard at work, but no explanation has been found yet.

---

[7] To find out more about tetraquarks, see for example http://www.quantumdiaries. org/2014/04/09/major-harvest-of-four-leaf-clover/

Hadron
any particle made of quarks
Meson
any particle made of one quark
and one antiquark
Baryon
any particle made of three quarks
or three antiquarks
Pion: up antidown positive pion
+ =
π^+
+2/3 +1/3 = +1
Kaon: up antistrange positive kaon
+ =
K^+
+2/3 +1/3 = +1
Proton: up up down proton
+ + =
+2/3 +2/3 −1/3 = +1
Neutron: up down down neutron
+ + =
+2/3 −1/3 −1/3 = 0

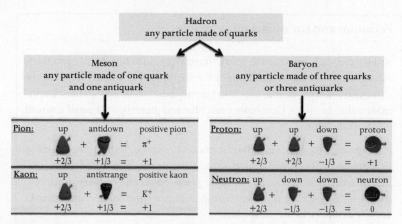

Figure 1.9 Any particle made of quarks is a hadron. The hadron family includes mesons, such as pions and kaons for example, and baryons, such as protons and neutrons.

Source: Pauline Gagnon and © Particle Zoo.

Pion: up antidown positive pion
+ =
π^+
+2/3 +1/3 = +1
Pion: antiup down negative pion
+ =
π^-
−2/3 −1/3 = −1
Kaon: up antistrange positive kaon
+ =
K^+
+2/3 +1/3 = +1
Kaon: antiup strange negative kaon
+ =
K^-
−2/3 −1/3 = −1

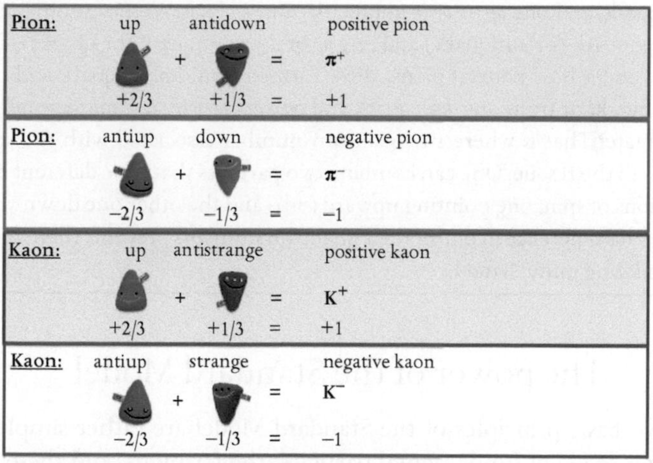

Figure 1.10 Mesons belong to the family of hadrons and contain one quark and one antiquark.

Source: Pauline Gagnon and © Particle Zoo.

Fermions and the exclusion principle

As I have already mentioned, quarks have a quantum number, or property, called *color*. Every flavor, each one of the six types of quark, in fact comes in three different colors: red, blue and green. These colors add up like the primary colors in optics. Combining red, blue and green light in equal amount yields white light.

Having different colors is handy: it makes it possible to have three quarks in a proton without violating the exclusion principle, which forbids having two identical quarks in the same place. Since one of their properties differs—the quantum number corresponding to their color—these quarks are allowed to be in the same place.

Only white, or neutral, combinations are allowed for hadrons. For a baryon, this is quite easy: one simply needs to combine three quarks of three different colors. For mesons, the particles made of one quark and one antiquark, we take one quark of one color and associate it with an antiquark having the complementary color. A red quark can combine with an antiquark of antired color. The most common mesons are the pions (made of one *up* or *down* quark, and one *up* or *down* antiquark) and the kaons (those containing a *strange* quark (or antiquark) and one *up* or *down* antiquark (or quark)). You may wonder how neutral pions, those formed from one *up* quark and one *antiup* quark, or from one *down* quark and one *antidown* quark, manage not to annihilate? That is where the quantum number associated with the spin comes to the rescue. One can combine two particles that have different orientations of spin, one pointing upward ($+\frac{1}{2}$) and the other one downward ($-\frac{1}{2}$). This difference in one of their quantum numbers prevents them from annihilating immediately.

The power of the Standard Model

The two basic principles of the Standard Model are rather simple: all matter is made of fundamental particles, the fermions, and these particles interact with each other by exchanging other particles, the force carriers known as bosons. This theory comes with a whole slew of complex equations, of course, enabling theorists to make extremely precise predictions.

The Standard Model establishes several relations between particles. It predicts the odds of producing various particles, as well as how often these particles break apart to give other particles. It also predicts in what proportions each possible decay type should occur. Some of its calculations have even been tested up to the ninth decimal point! It is an extremely powerful theory but, unfortunately, a flawed one, as we shall see in the sixth chapter. This is forcing theorists to seek a better and more encompassing theory. This still undefined theory should explain what is referred to as "new physics." I will return to this in the following chapters.

THE MAIN TAKE-HOME MESSAGE

The Standard Model tells us that all matter is made of fundamental particles, the twelve *leptons* and *quarks*. Each one has its own antiparticle, and all these particles belong to the *fermion* class. These grains of matter interact with each other by exchanging other particles called *bosons*. The particles in Figure 1.11 are the only known ones that are not made of something else; hence these are the fundamental particles.

By combining *up* and *down* quarks, we obtain protons and neutrons. These in turn form atomic nuclei. By adding electrons, we get the atoms and can form the 118 different chemical elements by varying the number of protons

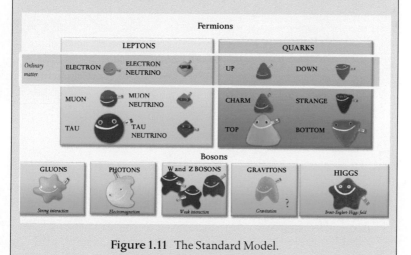

Figure 1.11 The Standard Model.
Source: Pauline Gagnon and © Particle Zoo.

continued

THE MAIN TAKE-HOME MESSAGE (*continued*)

in the nuclei. Hence, everything we see around us can be built from a construction set containing electrons and *up* and *down* quarks.

The two classes of particles, fermions and bosons, behave completely differently. We do not know why this distinction exists, nor why the fundamental particles are of such different masses or why they come in three generations. We are also clueless when it comes to explaining where all the antimatter produced after the Big Bang has gone. We do not understand either why the gravitational force is so much weaker than the other forces. In other words, we have a truly beautiful model that has greatly improved our understanding of the material world, but it still leaves us with several open questions.

2

What about the Higgs Boson?

The Standard Model, such as we know it today and as described in the first chapter, only appeared in 1967 when Abdus Salam and Steven Weinberg incorporated ideas developed during previous years by several people. Already in 1961, Sheldon Lee Glashow had managed to unify two of the fundamental forces described by the Standard Model. The electromagnetic and the weak forces have since then been described within the same theoretical framework under the name *electroweak force*.

Back in 1964, physicists were still completely unable to explain how the fundamental particles acquire their mass. The fragments of models that existed and their associated equations produced only massless particles, while it was well known that nearly all particles had mass (only photons and gluons have no mass). The importance of mass is enormous. It is a basic property of all fundamental particles. We also find it on our macroscopic scale, although, as we shall see, the mass of composite matter does not really come from the mass of its constituents.

In 1964, several theorists, whom we can see in Figure 2.1, were looking for a way to confer a mass on what was thought at the time to be a boson associated with the strong interaction. To do so, they developed a mathematical formalism known today under the name of the Brout–Englert–Higgs mechanism. This bears the names of the first three people who suggested it. These theorists were only adding their own contribution to a theory in the making, building on what had been developed by numerous predecessors. This is what Peter Higgs stressed with huge honesty in a presentation he gave in Stockholm in July 2013 during the largest conference in particle physics of the year, three months before receiving half of the Nobel Prize in Physics.

Figure 2.1 The theorists who contributed to what is now called the Brout–Englert–Higgs mechanism back in 1964: from left to right, Tom Kibble, Gerald Guralnik, Carl Hagen, François Englert, Robert Brout and Peter Higgs.
Source: Wikipedia and CERN.

So what is needed to confer a mass on fundamental particles? In fact, three different ingredients are needed. We are going to examine them in detail in the rest of this chapter. These three ingredients are:

1. The Brout–Englert–Higgs mechanism.
2. The Brout–Englert–Higgs field.
3. The Higgs boson.

So let's see how all this works.

The Brout-Englert-Higgs mechanism

When this mechanism was proposed in 1964, theorists were aiming simply to give a mass to some bosons. It was not until 1967 that Steven Weinberg, followed shortly by Abdus Salam, used the ideas of the Brout–Englert–Higgs mechanism to confer a mass on the Z and W bosons, as well as on leptons.[1] Later on, this mechanism was also applied to quarks. This mathematical formalism was therefore proposed before the birth of the Standard Model. Nowadays, it is used to reshape the equations of the Standard Model.

The electroweak force comes with four force carriers, four particles called bosons because they have integer spins, as described in the previous chapter. These are the photon, which is massless, and the $W+$, $W-$ and Z^0 bosons, which all have a mass. Without the Brout–Englert–Higgs mechanism, the equations describing these two combined forces yield four bosons, but none of them has a mass. Hence, the bosons coming

---

[1] As seen in the previous chapter, leptons correspond to a class of particles containing the electron, the muon, the tau and the three neutrinos.

Table 2.1 The four bosons of the electroweak force.

Boson	Mass in GeV	Electric charge
Photon	0	0
W^+	80.4	+1
W^-	80.4	−1
Z^0	91.2	0

out of the theory do not correspond to the real particles associated with these forces, since we know that three of them in fact have a mass.

But if we apply the Brout–Englert–Higgs mechanism to the Standard Model, it provides a way to give mass to some of them. This mechanism reshuffles the equations describing the electroweak force by simply adding small pieces of equations corresponding to a new field (which we will detail in the following sections). This clever trick allows us to rearrange the initial equations. Out of the new equations come four bosons as before, but this time three of them have a mass. The mechanism "breaks the initial symmetry." In other words, we start with four identical, massless bosons, and succeed in keeping one massless boson and obtain three massive bosons. These four bosons now correspond exactly to what one observes in nature: one massless boson, the photon, and three massive bosons, the $W+$, $W−$ and Z^0 bosons (Table 2.1).

The Brout–Englert–Higgs mechanism is much more than a simple trick to reshape the equations. It describes in mathematical terms a very real physical entity, known today as the Brout–Englert–Higgs field. The Brout–Englert–Higgs mechanism, which is necessary to break the symmetry in the equations of the Standard Model, thus reveals the existence of a new field.

The Brout–Englert–Higgs field

This is this field that confers mass on all particles, as we shall see. But what is a field? You may have heard of magnetic, electric and gravitational fields. All of these fields are invisible but their effects are perceptible. For example, a magnet generates a magnetic field. We perceive its effects when a magnet attracts metallic objects or when it pins down a small object on the refrigerator. But we can really visualize the lines of a magnetic field simply by sprinkling iron filings onto a piece of paper placed on a magnet, or onto the magnet itself, as can be seen in Figure 2.2.

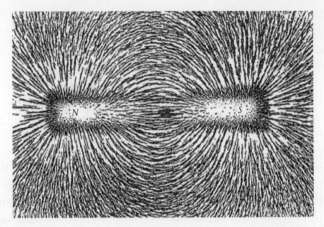

Figure 2.2 We can visualize the magnetic field of a small magnet simply by sprinkling it with iron filings.
Source: Wikipedia.

Similarly, an electric charge generates an electric field. Any other electric charge passing nearby will feel its effects and will move along the lines of the field. Massive celestial bodies, such as the Sun and the Earth, produce gravitational fields. We do not see them with our eyes, but we feel their effects. We are attracted toward the Earth, and every object that falls follows the gravitational field lines present there. These lines all point toward the center of the Earth.

These are some examples to illustrate the not-so-intuitive concept of a field. The Brout–Englert–Higgs field is similar to all the fields mentioned above except that there is no source. There is no equivalent of the magnet, the electric charge or the mass that generates the Brout–Englert–Higgs field. This field simply appeared almost immediately after the Big Bang[2] and has permeated the entire Universe ever since. This field is in fact a property of space, just as time and the three dimensions of space are properties of the world we live in. We can imagine it as the canvas of our Universe: that's simply the way our Universe was knitted.

---

[2] I will talk more about the Big Bang, the moment marking the first instant of the Universe, in Chapter 5. The Brout–Englert–Higgs field would have appeared one tenth of a billionth of a second after the Big Bang.

This field permeates all space around us. Without it, the fundamental particles would all move at the speed of light. But as soon as it is present, these particles interact with the field and are slowed down.

To describe better what happens, I need to introduce a few concepts. First, let's look at *energy conservation*, a fundamental principle in physics. This states that energy can take diverse forms but its total quantity is always preserved. Imagine that energy is a liquid: we could pour some of it into several different containers but the total quantity of liquid would stay the same. It would simply be distributed differently.

The second essential principle we need establishes the equivalence between mass and energy. This states that mass and energy are two forms of the same essence, just as two currencies both represent money but in different forms. The principle of equivalence between mass and energy is immortalized in the best-known equation in physics, and the only one to appear in this book: $E = mc^2$.

Here, E represents the energy and m represents the mass, or matter. The conversion factor between the two is c^2, the square of the speed of light. We can convert energy into mass (or matter) and vice versa. This works exactly as with two currencies: we can convert one into the other using the established exchange rate. The same thing happens here, except that the exchange rate, the square of the speed of light, is fixed.

Just as a large coin can be "broken" into smaller coins, a heavy particle can *decay* into smaller particles. The smaller particles are not initially contained in the larger particle, just as the smaller coins are not hidden inside the larger coin. Small coins can also be exchanged for a larger coin of equivalent value, and small particles can combine to give a larger particle. Sometimes, the larger particle is nevertheless a fundamental particle that is not made of any constituents. The equivalent in energy of all these particles is like the value of the coins: it is the entity that does not change.

A fundamental particle can have energy in two forms: either in the form of movement, it is then called *kinetic energy*, or in the form of mass. The mass of a particle can thus be seen as congealed energy.

My grandmother once told me that, because of age, she could hardly walk. She kept catching her feet on the carpet, and had the impression that she was tripping over the flowers depicted on it. The Brout–Englert–Higgs field plays a similar trick on fundamental particles. In the field's presence, they begin interacting with it. Instead of moving freely, they catch their feet in the "flowers" of the field and can no longer move freely.

One other way to visualize this is to imagine a person walking in an empty room: nothing hinders her progress. But if the same person has to cross a crowded room at a reception, she will have to make several stops to greet all her acquaintances. She will progress much less rapidly.

A particle without mass can move at the speed of light in a hypothetical space containing no Brout–Englert–Higgs field. All its energy is then in the form of movement, since it has no mass. Think again of the analogy of energy being like a liquid. We could then say that all of the energy of this particle is in a container labeled "movement," whereas the container labeled "mass" is empty (Figure 2.3).

Now, what happens to this particle if somebody "switches on" the Brout–Englert–Higgs field? It begins interacting with the field and can no longer move freely. It begins to catch its feet in the flowers on the carpet, just like my grandmother. However, this field has the peculiar feature that it does not cause the particle to lose energy. But since the particle moves more slowly, it has less kinetic energy. So, where has the missing energy gone? It is neither lost nor dissipated. This energy now simply appears in the form of mass (Figure 2.4).

The Brout–Englert–Higgs field does not cause any energy loss for fundamental particles. But since the particles move more slowly in its presence—they interact with it just like the person stopping to say hello—their kinetic energy is decreased. Their kinetic energy is not lost, but simply transformed into mass. In physics, mass is defined as the resistance to motion (or more specifically, the resistance to a change in motion). In the presence of the Brout–Englert–Higgs field, particles acquire a resistance to motion.

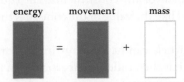

Figure 2.3 A fundamental particle carries energy in two forms: some comes from its movement (we call this kinetic energy), while the other part comes from its mass. If energy were a fluid, the left container, representing the total energy, could be split between the two other containers. This diagram depicts the situation for a massless particle: all its energy would be found in the form of movement.

Source: Pauline Gagnon.

Figure 2.4 In the presence of the Brout–Englert–Higgs field, a particle is slowed down by interacting with the field, as if getting entangled in it. Since it travels more slowly, less of its energy is associated with its movement. Its total energy remains unchanged but some of it now appears in the form of mass, as if some fluid had moved from the "movement container" to the "mass container." The particle is no longer massless; it has acquired mass.

Source: Pauline Gagnon.

How much mass does a particle acquire? The more it interacts with the Brout–Englert–Higgs field, the more mass it acquires. The better known a person is, the more she will interact with the crowd at a cocktail party and be slowed down. This applies to all particles: quarks, leptons and bosons. Since electrons interact very weakly with the field, their mass is very small. In contrast, the *top* quark has the highest interaction with the field and is thus the most massive particle.

The Higgs boson itself also acquires its mass from interacting with this field. On the other hand, photons do not interact with this field and thus remain massless. The question asked in the first chapter, "Why do fundamental particles have such a broad range of masses?" could be rephrased as "Why do particles interact so differently with the Brout–Englert–Higgs field?" There is no answer to this second question either. It still remains a mystery.

How much mass does each particle acquire?

The more strongly a particle interacts with the Brout–Englert–Higgs field, the more mass it acquires. This purely theoretical assertion made by Brout, Englert and Higgs has now been confirmed experimentally. This is illustrated in Figure 2.5 by preliminary results obtained by the CMS Collaboration. This group of researchers has checked this assertion by measuring how often Higgs bosons decay into given types of particles. This is called the *coupling* and represents the intensity of the interaction of a particle with the

continued

How much mass does each particle acquire? (*continued*)

Brout–Englert–Higgs field. The vertical axis gives the value of the coupling for each particle, the mass of which is given on the horizontal axis.

Note the use of logarithmic scales on the two axes, which cover several orders of magnitude. We see, for example, the experimental measurements for the tau lepton (denoted by the symbol τ), the *bottom* quark (*b*), the *W* and *Z* bosons and, finally, the *top* quark (*t*). The vertical lines give the experimental uncertainty in the measurements. The red line gives the value of the *standard deviation*, that is, the experimental uncertainty that corresponds to a reliable confidence level of 68%. There is a 68% chance that the real value lies somewhere within this interval. The blue line corresponds to two standard deviations, with a confidence level of 95%. The shaded areas in yellow and green indicate zones that have confidence levels of 68% and 95%, respectively, when all the measurements are compared with the theoretical predictions, taking into account the individual error margins for each particle. We see that the current measurements all agree with the theoretical predictions of the Standard Model, denoted by "SM Higgs" and given by the dotted line.

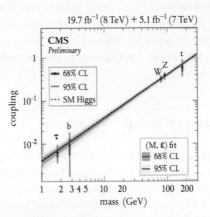

Figure 2.5 The mass of a particle depends on how strongly it interacts with the Brout–Englert–Higgs field. The more it interacts, the heavier it is.
Source: CMS.

What gives mass to composite matter?

Another essential point must be clarified: the Brout–Englert–Higgs field confers mass on all fundamental particles, but not on composite matter, contrary to what is frequently thought. Here is why. In an atom, the mass comes essentially from the atomic nucleus, the electron being approximately 1840 times lighter than protons and neutrons. For particles, one measures mass in MeV, a unit of energy, since as we have seen, mass and energy are equivalent.[3] The symbol MeV represents a megaelectronvolt, that is, a million electronvolts. The electronvolt is the energy acquired by an electron when it is accelerated across a potential difference of one volt.

The mass of a proton is 938 MeV. As we saw in the previous chapter, a proton is made of three quarks and some massless gluons that bind them together. But the sum of the masses of these three quarks is a mere 11 MeV, barely one percent of its total mass. Suppose you were to put three balls in a bag, each one weighing about 4 grams. Then, imagine your surprise if you weigh the bag and the balance indicates 938 grams!

Once again, the principle of equivalence between mass and energy comes into play to explain how the nucleons (as protons and neutrons are generically called) get their mass. A nucleon mass comes from the energy of motion of its three quarks and from the energy carried by the gluons (Figure 2.6). The quarks are free to move within a nucleon. This is due to the fact that the strong force is weak at very short distance and becomes larger only if the quarks try to move away from each other. This explains why the quarks are confined within nucleons, a phenomenon called *asymptotic freedom*. Nevertheless, the strong force only acts on short distances, barely larger than the size of a nucleon.

Thus 99% of a proton or neutron mass comes from the energy of its constituents. Likewise, a nucleus has a lot of *binding energy* that keeps all of its nucleons tied within the nucleus. This is the energy source in a nuclear reactor. When a nucleus is split, the bonds between protons and neutrons are broken and this energy is released.

---

[3] Strictly speaking, one should express the mass in units of MeV/c^2 but, to simplify things even further, physicists use a system of units where the speed of light, c, is taken to be 1.

- Mass of quarks: 11 MeV
- Mass of proton: 938 MeV

Figure 2.6 Taken together, the masses of the three quarks inside a proton account for only 11 MeV. Most of a proton mass comes from the energy of motion of its three quarks and from the energy carried by the gluons that keeps them confined together. This energy constitutes most of the mass of the proton, namely 938 MeV. The Brout–Englert–Higgs mechanism confers mass only on fundamental particles, not on composite matter such as protons, neutrons and atoms. Likewise, the mass of atoms comes from the binding energy between the constituents in the nucleus.

Source: Wikipedia and Pauline Gagnon.

What about the Higgs boson?

Now that the role of the Brout–Englert–Higgs field has been clarified, we can finally talk about the Higgs boson. First of all, we should mention that this boson was named in honor of Peter Higgs not because he was the first person to publish an article on the Brout–Englert–Higgs mechanism but, paradoxically, because his article was initially rejected. François Englert (Figure 2.7) and Robert Brout published their work first, beating Peter Higgs by a good month and another team of three theorists, Tom Kibble, Gerald Guralnik and Carl Hagen, by several months. These three groups had been working independently on the same topic. An editor had refused Peter Higgs' article because of a lack of concrete predictions, and had consequently declared that article to be of no scientific interest. Peter Higgs resubmitted his article somewhere else, taking good care this time to mention that the mechanism he proposed implied the existence of a new boson. He was thus the first person to mention the existence of a new particle explicitly. The particle could reasonably be renamed the "scalar boson" or the "H boson." However, the fame of the "Higgs boson" is now such that it is highly unlikely that its name will ever change. This twist of fate led Peter Higgs a few years ago to publish an article entitled "My life as a boson"![4]

So, what about this famous Higgs boson, then? In fact, it can be seen simply as an excitation of the Brout–Englert–Higgs field. If we compare

[4] http://www.worldscientific.com/doi/abs/10.1142/S0217751X02013046

Figure 2.7 Professor François Englert standing in front of the ATLAS detector during a visit to CERN in December 2007.
Source: CERN.

the field to the surface of the ocean, the Higgs boson would be a wave on the surface of this ocean. To excite the ocean and produce waves, it is enough to supply energy, which might be from wind, tides or an earthquake. The same thing goes for the Brout–Englert–Higgs field: one can excite it by supplying it with energy, something particle accelerators can do. This excitation is nothing but the Higgs boson itself.

Disappointing? Not really. Imagine that I have in front of me a small aquarium made from panes of glass. If I claim that it is filled with water, I can prove my claim simply by tapping a pane with my hand. If it is indeed full of water, wavelets will appear on the surface of the water. No water, no waves. And no Higgs boson without the Brout–Englert–Higgs field.

Theorists had suggested that the Universe was filled with the Brout–Englert–Higgs field. Experimental physicists proved that a field was indeed there by producing excitations of this field in the form of Higgs bosons. At the LHC at CERN, the LHC supplied the energy necessary to excite the field. The discovery of the Higgs boson proved the existence of the field that gives mass to fundamental particles. In the next chapter, we shall see what the LHC is, how it works and how the ATLAS and CMS detectors (Figure 2.8) revealed the presence of the Higgs boson.

Before ending, I would like to settle the question of the expression "God particle," a joke that is as worn out as it is resilient. This expression comes

Figure 2.8 Even Professor Peter Higgs seems to have a hard time believing his eyes during his visit to the CMS detector in 2008.
Source: CERN.

from a publisher who turned his puritanism into a successful marketing operation. Leon Lederman, an American physicist well known for his good sense of humor and one of the laureates of the Nobel Prize in Physics in 1988, frustrated after not having found the Higgs boson after nearly 30 years of research, suggested entitling his popular science book *The God-damn Particle*. But his publisher refused, judging the title inappropriate. Instead, he suggested *The God Particle*. Hence Lederman's book appeared in 1993 under the title of *The God Particle: If the Universe Is the Answer, What Is the Question?*[5] Unfortunately the name endures even today, creating useless confusion about an already complex topic. The name "God particle" means absolutely nothing, and it is high time to move on. Granted, it was

⁵ Leon M. Lederman and Dick Teresi, Dell Publishing, 1993.

the last missing particle needed to complete the Standard Model picture, an essential and very special particle, but there is no need to exaggerate.

THE MAIN TAKE-HOME MESSAGE

Until 1964, the equations of what was to become the Standard Model predicted only massless particles, in contradiction to experimental observations. Several theorists then proposed the existence of a field that fills the entire Universe. Fundamental particles get their mass by interacting with this field, called the Brout–Englert–Higgs field and described in mathematical terms by a mechanism bearing the same name. This field is a very real but invisible physical entity. It hinders the propagation of fundamental particles and acts similarly to a crowd filling a room. A person will have difficulty crossing this crowded room without stopping to greet various acquaintances. The crowd will slow her down considerably, as if the person had become more massive. For a fundamental particle, this slowing down corresponds to converting some of its kinetic energy (associated with its movement) into mass. This takes place without energy loss. This comes from the equivalence principle that stipulates that mass and energy are two forms of the same essence and from the principle of conservation of energy. No energy is ever lost, just transformed. The Higgs boson is an excitation of the Brout–Englert–Higgs field, just as a wave is an excitation of the surface of the ocean. Its existence proves the presence of this field. However, this field only gives mass to fundamental particles. Composite objects of matter such as neutrons and protons, which carry most of the mass of atoms and, hence, of all matter, get their mass mostly from the energy carried by the quarks, as they move around, and the gluons that bind the quarks together. Their mass can therefore be seen as congealed energy. This follows from the equivalence between mass and energy expressed by the equation $E = mc^2$.

3

Accelerators and Detectors: The Essential Tools

Finding a Higgs boson is no small feat. First of all, it needs to be produced before we can find it. This is done by concentrating large amounts of energy into a tiny point of space to "excite" or generate a "wave" in the Brout–Englert–Higgs field, as we saw in the previous chapter. The only machines on Earth powerful enough to produce Higgs bosons are particle accelerators such as the Large Hadron Collider[1] at CERN (Figure 3.1). Elsewhere in the Universe, Higgs bosons are very likely produced when protons of very high energy coming from cosmic rays collide with protons or neutrons in the upper atmosphere or even on the surface of the Moon. And, who knows, there could also be extra-terrestrial civilizations out there that are equipped with accelerators as powerful as the LHC.

The Large Hadron Collider got its name because it brings protons, which are particles made of quarks, into collision. Protons belong to the hadron family, hence the name. And, as its name also suggests, the Large Hadron Collider is big, gigantic even. It deserves all sorts of superlatives: the biggest, the most powerful, the most successful, the coldest, the most . . . everything. It is really impressive.

Like all particle accelerators, it generates locally enormous quantities of energy from collisions of particles to create, or rather produce,[2] new particles. Once again, the equivalence principle between mass and energy comes into play to transform pure energy into matter. To do this, the LHC accelerates heavy particles, most of the time protons, to near

---

[1] Until September 2011 the Tevatron, an accelerator located near Chicago, was also powerful enough to produce Higgs bosons but could not produce them in sufficient numbers to allow discovery of the particle. It has since ceased to operate.

[2] During Pope John Paul II's visit to CERN some 30 years ago, his guide spoke about the creation of particles in the collision of particle beams. The Pope corrected him, saying "You mean *production; creation* is my business!"

Figure 3.1 The 17 mile (27 km) ring of the Large Hadron Collider drawn on top of an aerial photograph of the site. In the background, one can see Lake Geneva and the city of Geneva, and, further away, the Alps. The accelerator is actually located 300 ft (100 m) underground.

Source: CERN.

the speed of light, that is, about 186,000 miles (300,000 kilometers) per second. The protons circulate in two parallel tubes, forming beams that are brought into collision at four different points, at the center of four detectors. The energy released during these collisions materializes in the form of diverse particles. The more the energy, the heavier the particles that can be produced (the more money one has, the bigger the car one can afford). These particles are very unstable and break up almost immediately into several fragments, namely more stable particles. The role of the detectors is to catch all these fragments and reconstruct the particles that were initially produced. They act like giant cameras, taking photos of these miniature explosions and reconstructing the initial particles from the debris.

In summary, we have an accelerator (Figure 3.2) that accelerates protons to produce new particles from the energy released during

Figure 3.2 A portion of the 17 mile (27 km) ring of the LHC, with some of its 1232 superconducting dipole magnets.
Source: CERN.

collisions, and detectors to detect the particles that are produced. Let us see in detail how all this works.

The Large Hadron Collider

Thirty-eight thousand tons of high technology, combining gigantism and extreme precision, the LHC is located in a 17 mile (27 km) long tunnel that was built for CERN's previous accelerator, the Large Electron–Positron Collider or LEP. The LHC accelerator consists of 1232 dipole magnets and 392 quadrupole magnets, plus a handful of more complex magnets, all superconducting and operated at −456.3 °F (−271.3 °C), a mere 3.4 °F (1.9 °C) above absolute zero (Figures 3.3 and 3.4). Superconductors, as we saw in Chapter 1, offer no resistance to the passage of an electric current. Certain materials, such as the niobium–titanium alloy used for the LHC magnets, become superconducting when they are cooled to very low temperatures. Superconductivity yields much more powerful magnets than do ordinary conductors since they can sustain larger currents. The LHC magnets carry a current of 12,000 amperes, a thousand times what is customarily found in a household circuit.

Conventional magnets would not have been powerful enough to bend the beams and maintain them in their circular orbit in the accelerator: a 75 mile (120 km) ring would have been needed. The machine is gigantic enough as it stands; any bigger would have been prohibitive.

An electrically charged particle beam can be manipulated with magnets, just like a beam of light can be deviated with prisms and lenses.

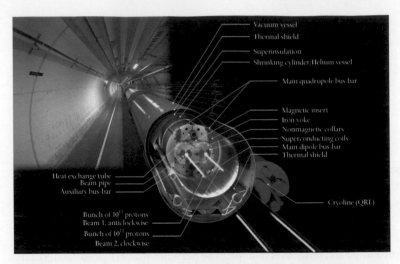

Figure 3.3 A schematic view of a dipole magnet showing its key components.
Source: CERN.

Figure 3.4 Installation of one of the 1232 LHC dipole magnets in the tunnel
300 ft (100 m) underground.
Source: CERN.

Dipole magnets are used to bend the trajectories of the protons and maintain them in a circular orbit, whereas *quadrupole* magnets are used to focus the beams. In other words, these magnets squeeze the beams. Other *multipole* magnets add various corrections to the proton trajectories. It is absolutely necessary to keep all of the protons in tow to maintain them in orbit for several hours, knowing that each second, the protons go around the 17 miles of the LHC 11,245 times.

In total, 4750 miles (7600 km) of cables, each one containing 250,000 strands, were needed for the magnet windings. The total length of the strands is equivalent to the length of six round trips to the Sun, plus 136 round trips to the Moon and 24 Montreal–Paris flights, with enough left over for 1046 visits to the corner shop. It is not so surprising, in view of all this, that the construction of the machine took 15 years, especially since some of the technologies needed did not exist at the start of the project and had to be invented along the way.

Jumping ahead of one's time

For example, the availability and cost of the computing power and storage capacity needed for the entire LHC project (accelerator and detectors) were estimated using Moore's law by extrapolating from existing technology. Moore's law states that every year or two, computers twice as powerful and storage capacity twice as large become available for the same price. Likewise, the physicists designing the trigger and data acquisition systems[3] banked on being able to use a new generation of faster electronic modules before they existed, to meet the needs of their experiments.

For the LHC machine itself, the first papers on its design appeared in the mid 1980s. The scientists and engineers involved estimated then that the best performance reached in prototypes by superconducting magnets at that time would be achievable on the industrial scale needed by the LHC (several thousands of magnets) within 10 years or so. And this was indeed accomplished![4] The same goes for the technology required to connect superconducting cables. All aspects of the technology needed, such as inductive soldering and joining by ultrasound, existed

[3] I describe these systems briefly later in this chapter.
[4] For example, in the 1980s, superconducting magnets could carry 2000 amperes per square millimeter at 4.2 K with a magnetic field of 5 tesla. The LHC magnets carry 50% more current, namely 3000 amperes per square millimeter, under the same conditions.

in other domains, but teams at CERN, in collaboration with other laboratories and several industrial partners, developed machines able to meet the strict specifications and scale of the LHC project. This work started in the late 1990s and the results were ready for use in the LHC tunnel in 2005. Similarly, welding and cutting machines for superconducting lines existed, but they also had to be adapted to the specific configuration of the LHC tunnel. To this day, the LHC is still the largest and coldest cryogenic installation ever achieved on such a scale.

A very special ring

The huge LHC ring was built 300 ft (100 m) underground for two reasons. Firstly, it was absolutely necessary to shield the detectors from cosmic rays, since these would interfere with the measurements. Second, it was also essential to protect people and the environment from radiation. Building at the surface was also unthinkable given the cost of real estate in the area.

The two proton beams circulate in two separate vacuum pipes from which all air has been evacuated (Figure 3.5). Otherwise, collisions with

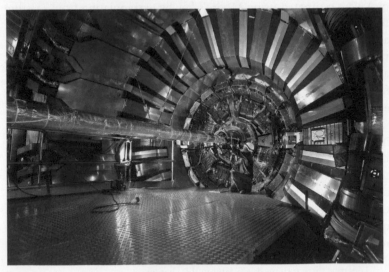

Figure 3.5 The accelerator beam pipes reach into the heart of each of the four LHC detectors, as here in the CMS detector.

Source: CERN.

air molecules would prevent the protons from traveling more than about 1/32 of an inch (1 millimeter). Powerful vacuum pumps maintain the pressure at 10^{-10} millibar, that is, 10^{13} or 10,000,000,000,000 times less than atmospheric pressure. In other words, there are 10^{13} times fewer air molecules per unit volume, for example per cubic inch or cubic centimeter, in the LHC pipes than in the air that we inhale. A special material invented at CERN, called a "getter," coats the beam pipe walls. Once warmed, this material absorbs the remaining molecules left by the vacuum pumps and acts like the sticky paper strips used as flycatchers. Of course, the beam pipe must be perfectly tight. A car tire as leaktight as the LHC beam pipe would take a million years to deflate.

The LHC is both huge and extremely sensitive to the slightest perturbation. For example, we know that the Moon's gravitational pull creates tides. This is usually only observed in large volumes of water such as the oceans and not in the Earth's crust, since the Earth's crust is much less fluid. However, the Earth's crust too undergoes a tiny deformation twice daily, due to the attraction of the Moon, but this is hardly perceptible. Nevertheless, this action of the Moon requires the LHC operators to constantly correct the proton trajectories to maintain them inside the LHC beam pipe, since it moves with the Earth's crust. One could say that the LHC has confirmed the presence of the Moon, although it was not exactly built for that purpose.

The CERN accelerator complex

Where do all the protons used by the accelerator come from? The great adventure of the LHC begins with a simple hydrogen bottle that provides these protons. Hydrogen is the simplest chemical element. Its nucleus contains a single proton, around which an electron revolves. The hydrogen atoms are stripped of their electrons by means of an electric field (Figure 3.6). The resulting protons are then accelerated by another strong electric field inside a small linear accelerator called Linac 2. They emerge from it with an energy of 50 MeV, that is, 50 million electronvolts. At this point, they are already traveling at one third of the speed of light. An electronvolt is the energy acquired by an electron subjected to a potential difference of 1 volt. An electron accelerated between the two poles of a 1.5 volt battery thus gains 1.5 electronvolts of energy. The energy supplied by the Linac 2 accelerator corresponds to the energy

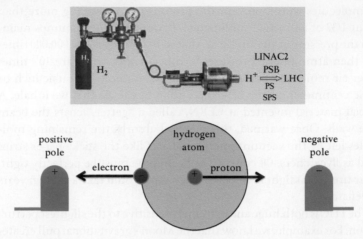

Figure 3.6 The protons for the beams are extracted from hydrogen atoms that have been stripped of their electrons using an electric field.

Source: Pauline Gagnon/CERN.

that would be gained if we applied a potential difference of 50 million volts! The protons leave Linac 2 and get an energy boost in the Synchrotron Injector (also called the Booster), a small circular accelerator that allows them to reach 1.4 GeV (giga-eV, that is, one billion electronvolts or one thousand MeV).

The next stage is the Proton Synchrotron, or PS, a circular and synchronized type of accelerator. This works roughly like a battery where the two poles are constantly switched. Essentially, a negative pole attracts the protons to the location of the pole. Then, as soon as they reach this point, the polarity of the pole is inverted (i.e., it becomes positive). This positive pole then repels them further toward the next negative pole, allowing them to keep their momentum. This inversion of polarity is synchronized with the passage of bunches of protons. The PS is the oldest accelerator still in operation at CERN. A circular accelerator has the advantage of being able to inject a little more energy into protons at every revolution, whereas a linear accelerator has only a single opportunity. The protons emerge from the PS with an energy of 25 GeV ready for the next stage, the Super Proton Synchrotron or SPS, the same type of accelerator as the PS but 11 times bigger (Figure 3.7). There they reach an energy of 450 GeV.

Figure 3.7 The Super Proton Synchrotron, or SPS, the third stage of the accelerator complex leading to the Large Hadron Collider.
Source: CERN.

At long last, the protons are injected into the LHC (Figure 3.8), where they undergo their final acceleration stage. In 2010, they reached an energy of 3.5 TeV, or 3.5 teraelectronvolts, that is, 3500 GeV, in about twenty minutes. This was increased to 4 TeV in 2012 and, in 2015, after 2 years of maintenance and consolidation work, the protons reached 6.5 TeV. The energy of the collisions corresponds to twice the energy of each beam. The energy available was thus 8 TeV in 2012 and reached 13 TeV on May 20, 2015. This increase in energy not only allows the production of heavier particles, opening the door to new discoveries, but also produces them more copiously, facilitating those discoveries.

In one year, the CERN accelerator complex barely uses one microgram of hydrogen to produce all the protons it needs, but it consumes huge amounts of electricity: 1260 GWh (gigawatt-hours, that is, 1000 megawatt-hours) per year when all the accelerators are in operation. This is equivalent to one and a half times the electricity produced in an average nuclear power plant in the United States.

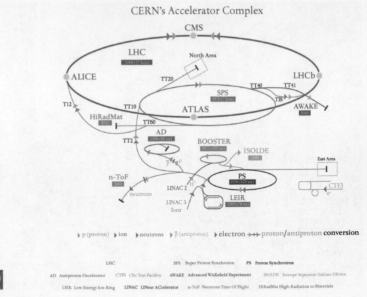

Figure 3.8 The chain of accelerators at CERN: equipment from previous decades is reused to feed the latest and most powerful accelerators.

Source: CERN.

Table 3.1 Accelerators in the CERN accelerator complex.

Accelerator (year built)	Size	Energy	Percentage of speed of light
Linac (1972)	111 ft (34 m)	50 MeV	31.4%
Booster (1972)	492 ft (150 m)	1.4 GeV	91.6%
PS (1959)	2060 ft (628 m)	25 GeV	99.93%
SPS (1973)	4.375 mi (7 km)	450 TeV	99.9998%
LHC (2008)	16.9 mi (27 km)	8 TeV	99.9999993%
LHC (2015)	16.9 mi (27 km)	13 TeV	99.9999997%

Is 1 TeV a lot of energy?

The LHC supplies an energy of 8 TeV, or 8 teraelectronvolts, to every proton it accelerates. The prefix *tera* represents a multiple of 10^{12}, or 12 orders of magnitude. A TeV is thus worth 1,000,000,000,000, or one million million, electronvolts. Is this a lot? It is equivalent to the energy of a 2 milligram mosquito in full flight, given that a mosquito can easily reach a speed of 1.4 km/hour, that is, nearly 1 mile an hour. On our own scale, this is extremely small but for a proton, which is less than a *fermi* (10^{-12} mm, or a millionth of a millionth of one millimeter) in size, it is enormous. Comparing the size of a proton to that of a 5 mm mosquito is like comparing the size of the same mosquito to the distance from the Earth to the Sun. Imagine now that all our mosquito's energy is condensed into the size of a proton, and it becomes gigantic.

We can measure the energy of this mosquito in the following way: its energy is given by ½ × mass × square of the speed, or $\frac{1}{2}mv^2$. Let us convert all the units to kilograms and meters per second to obtain the energy in joules (the joule is the unit of energy corresponding to $1 \text{ kg} \times 1 \text{ m}^2/\text{s}^2$). The mass of the mosquito is 2 mg, that is, 2×10^{-6} kg, and its speed is 1400 meters in 3600 seconds. That is approximately 0.4 meter per second. Its energy is thus ½ × 2×10^{-6} kg × $(0.4 \text{ m/s})^2$, or 1.6×10^{-7} joules. If we now convert these joules to electronvolts, knowing that 1 joule is worth 6.24×10^{18} electronvolts, we find that our 2 mg mosquito flying at 1.4 km/h has an energy of about 1×10^{12} electronvolts, that is, 1 TeV. Keep in mind that the electronvolt is a tiny quantity of energy on our macroscopic scale since it is defined by the energy associated with an extremely small subatomic particle, the electron.

Four big detectors

There are four big detectors around the LHC: ATLAS, ALICE, CMS and LHCb. Each detector was built by a "collaboration," that is, a group of physicists coming from hundreds of research institutes in various countries. Thousands of physicists, hundreds of engineers and thousands of technicians contributed to the building of each detector, using different technologies, materials and principles.

The LHCb Collaboration aims to discover where all the antimatter that was produced shortly after the Big Bang has gone, leaving no visible trace today. They are studying *b* quarks, hoping to unveil different behavior for matter and antimatter. The ALICE team is concentrating on the study of a state of matter that existed just an instant after the Big Bang, a state called the *quark–gluon plasma* (see box). The CMS and ATLAS detectors are multipurpose experiments with very broad research programs that include but are not limited to the Higgs boson, supersymmetry, dark matter and physics beyond the Standard Model. Both teams are pursuing the same types of research, including the fields of activity of ALICE and LHCb, but without being so specialized. One group's findings can thus be cross-checked by at least one other team.

Table 3.2 The four big detectors around the LHC.

	ATLAS	ALICE	CMS	LHCb
Height	82 ft (25 m)	53 ft (16 m)	49 ft (15 m)	33 ft (10 m)
Length	148 ft (45 m)	85 ft (26 m)	69 ft (21 m)	69 ft (21 m)
Weight	7000 tons	10,000 tons	14,000 tons	5600 tons
Number of scientists	3000	1000	3000	700
Number of institutes	177	100	179	65
Number of countries	38	30	41	16
Research	Multipurpose	Quark–gluon plasma	Multipurpose	Antimatter and *b* quark

---

The quark–gluon plasma: as hot and as cool as it gets

The difference between the three most common states of matter—solid, liquid and gas—is determined more or less by the amount of freedom of its molecules. The molecules of a gas are less bound than those of a liquid or a solid. A *plasma* is another state of matter, where atoms are so overexcited that they break apart. There is matter in the form of plasma in the Sun, but it can also be found in a flame or in a neon tube. The voltage applied to a

The quark–gluon plasma: as hot and as cool as it gets (*continued*)

neon tube supplies the energy needed for electrons to break free from the atoms, letting a cloud of electrons float around positive ions. A quark–gluon plasma is a state that is excited a notch higher. There is so much energy at play that even the nuclei, the protons and the neutrons, break apart, releasing their quarks and gluons, letting them coexist in a superenergetic soup.

The quark–gluon plasma existed only at the very beginning of the Universe, a mere tenth of one billionth of a second after the Big Bang. As the Universe expanded, it started cooling down, allowing the quarks and gluons to slow down enough to form protons, neutrons and other hadrons. The quark–gluon plasma state reappeared for the first time in the Universe 13.8 billion years later in the Super Proton Synchrotron at CERN in 2000, then at Brookhaven Laboratory in the United States, and now at the LHC. The collisions that occur in the LHC generate temperatures 100,000 times higher than those prevailing in the center of the Sun.

The LHC can accelerate not only protons, but also heavy ions such as lead atoms stripped of their electrons. These nuclei contain 82 protons and 125 neutrons. For approximately one month each year, the protons in the LHC are replaced by lead ions, causing even more energetic collisions, enough in fact to produce some quark–gluon plasma (Figure 3.9). The plasma, even

Figure 3.9 Simulation of a collision between two lead nuclei as it would appear just after the impact. Each nucleus contains 82 protons and 125 neutrons. The quarks, which are initially contained in the protons and neutrons, are shown in red, green and blue, whereas the still intact protons and neutrons appear in white.
Source: CERN.

continued

The quark–gluon plasma: as hot and as cool as it gets (*continued*)
....

though it is made of separate particles, exhibits group behavior, just as a swarm of bees moves with cohesion and fluidity. It is in fact a superfluid, that is, a fluid that has no viscosity. The viscosity of a substance determines whether it is as sticky as honey or as fluid as water. A superfluid is so fluid that it cannot be confined; it spreads in all directions and even flows out of its container. Hard to find cooler than that!

The collisions

In the LHC beams, the protons were grouped most of the time into 1404 bunches, each one containing one hundred billion protons (give or take a few). Every 50 billionths of a second (nanoseconds), a proton bunch met another proton bunch from the other beam traveling in the opposite direction. At the end of 2015, this rate doubled to produce yet more collisions, with 2808 bunches colliding every 25 nanoseconds. These collisions are made to happen at the dead center of each of the four detectors (Figure 3.10). Only a handful of these protons will collide at each

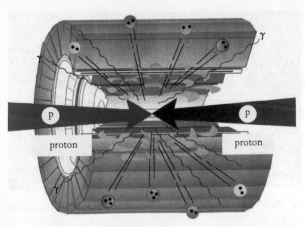

Figure 3.10 Schematic representation of two proton beams circulating in opposite directions around the accelerator and colliding in the center of a detector.

Source: CERN.

crossing, releasing the energy needed to produce all sorts of particles, from the best known to the rarest.

The ones easiest to produce are necessarily the best-studied ones, since we have had them in abundance for a long time. These particles supply us with precious landmarks that allow us to calibrate our instruments. But all these particles, whose discoveries were once rewarded with Nobel Prizes, can be a nuisance today. They form a background noise that masks the new, undiscovered phenomena. We thus have no choice: to detect particles produced far less often, we must plow through thousands of billions of these run-of-the-mill types of events to reveal new particles. However, knowing exactly how much background noise is expected is essential for us to be able to confirm the presence of any excess. Hence, we must collect billions of events and sort them to extract those possessing the uncommon characteristics of the particles that we want to study.

The energy released during such collisions between two protons materializes in the form of particles. Heavy and short-lived, they break apart nearly immediately into a multitude of fragments. Every collision looks like a miniature firework. One must catch all the fragments to reconstruct the initial particle. This is what we call a *particle decay*, and it is very similar to making change for a coin (Figure 3.11), as we have seen in Chapter 2. A large-value coin can be exchanged for smaller ones. A heavy, unstable particle (such as a Higgs boson) breaks apart into lighter particles almost immediately after it is produced. The energy equivalent to its mass reappears in the form of lighter, more stable particles. The purpose of the detector is to determine the origin, trajectory, direction,

Figure 3.11 The decay of a heavy, unstable particle is similar to making change for a large coin. Here is one way to break up a euro coin. The smaller coins were not inside the one-euro coin, but their combined value is one euro. Likewise, the energy equivalent to the mass of a heavy, unstable particle reappears in the form of lighter, more stable particles.

Source: Pauline Gagnon, Pixabay.

energy, electrical charge and identity of each of these particles to determine what particle was initially produced.

The detectors

Of the four LHC detectors, CMS (Figure 3.12) is the heaviest, with a weight of 15,000 tons (14,000 metric tons), twice as much as the Eiffel Tower. The ATLAS detector is the largest. In volume, it is about half the size of the Notre-Dame Cathedral in Paris. But, unlike these two huge monuments, particle detectors are made of hundreds of millions of small, ultraprecise parts, each one fabricated and assembled by hand. In the movie *Particle Fever*, my colleague Monica Dunford compares the ATLAS detector to a gigantic Swiss watch. The four LHC detectors indeed combine gigantism and high precision.

A detector is made of several concentric layers, exactly like a set of Russian dolls or the layers of an onion. Each layer is designed to collect part of the information and has the shape of a tin can: a central cylinder and two lids. The whole must be perfectly hermetically sealed: no particle crossing it should escape detection.

Figure 3.12 The CMS detector in all its glory.
Source: CERN.

From now on, I will describe the ATLAS detector (Figure 3.13), since this is the one I know best, having participated in its design, construction and operation. The other detectors are similar but use different technologies. This is essential. The observation of a new phenomenon will leave no room for doubt if two completely different instruments make the same observation. We shall see in the next chapter that this was the case for the Higgs boson.

There are always several ways to make the same measurement, as in this story about a physics professor who asked a student to estimate the height of a building using a barometer. Of course, the student knew that all she had to do was to measure the difference in pressure between ground level and the roof of the building to estimate its height. But she felt that that was a rather convoluted way of measuring the height of a building. So she suggested instead that one might as well make a pendulum from the barometer and determine the height of the building from the oscillation frequency of the pendulum. When her professor rejected this answer, she proposed a second method: throw the barometer from the roof of the building and measure the time that elapses before it crashes to the ground. As her professor was getting impatient, she finally suggested that she could simply offer the barometer to the caretaker of the building in exchange for the information.

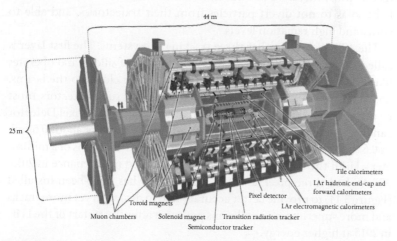

Figure 3.13 A schematic view of the ATLAS detector, one of the four gigantic detectors operating at the Large Hadron Collider.
Source: ATLAS.

For every measurement, however, there is only one good answer and the same result must be obtained no matter what method is used. For particle detectors, several subdetectors are needed to extract the various pieces of information. Several technologies exist for each subsystem. The ATLAS and CMS Collaborations chose different methods, thereby increasing the reliability of any claim about a new observation. Roughly speaking, each detector needs four different layers: a tracking system to reconstruct the trajectories of charged particles, calorimeters to determine the energy of every particle, magnets to provide a magnetic field that allows the trackers to determine the charge and momentum of charged particles, and muon detectors—I shall let you guess what those are used for.

The tracking system

The role of the tracking system is to reconstruct the trajectories of all electrically charged particles. Neutral particles do not leave a track in these detectors. The trackers are placed as close to the beams as possible—less than 4 inches (10 cm)—so that they are able to reconstruct as exactly as possible the point of origin of every track. But the closer one gets to the beams, the more intense the radiation is. These detectors must therefore be built from materials that are both extremely light so as to not divert particles from their trajectories, and able to withstand high radiation levels.

The ATLAS detector has three tracking subsystems. The first layer is called the Pixel Detector. This detector is made of silicon and operates like a digital camera. The track density is higher closer to the beams. So, to be able to distinguish the different tracks, such detectors must be ultraprecise. With its 80 million channels, the ATLAS Pixel Detector can determine the position of a track with a precision of about 0.6 mil, or 0.6 thousandth of an inch (14 microns, or 14 thousandths of a millimeter). Until the first technical stop for scheduled maintenance in 2013, ATLAS had three pixel layers. A fourth layer has since been installed (Figure 3.14) to provide better accuracy and to deal with yet more tracks and more superimposed collisions per event after the restart of the LHC in 2015 at higher energy.

When proton bunches cross, several protons collide simultaneously. The extra pixel layer now provides higher precision when reconstructing the origin of the tracks. Most of these collisions correspond to cases

Figure 3.14 Insertion of the fourth Pixel Detector layer into the heart of the ATLAS detector in May 2014.
Source: Heinz Pernegger, ATLAS.

where protons barely graze each other. These events carry little energy. Only head-on collisions are energetic enough to produce something interesting. One must associate every single track coming out of the collision with a precise collision point, in order to retain only tracks coming from the same collision point, as can be seen in Figure 3.15. Here, 25 minor collisions took place at the same time as a very energetic one. We can see clearly the 25 separate collision points in the enlarged section underneath. The two very energetic tracks shown by the bright lines (in yellow) emerge from the same collision point. No other tracks are associated with that collision point. We can thus ignore all the other tracks coming from low-energy collisions and simply retain these two tracks.

Moving outward from the collision point, the next layer is the Silicon Tracker, another semiconductor tracker. Its purpose is also to detect the passage of every charged particle with high precision. This detector is made of microscopic silicon strips that are activated by the passage of a charged particle. It contains two double concentric layers. Every charged particle thus leaves four dots on its passage, and these dots, once connected, reconstruct the trajectory of the passing charged particle.

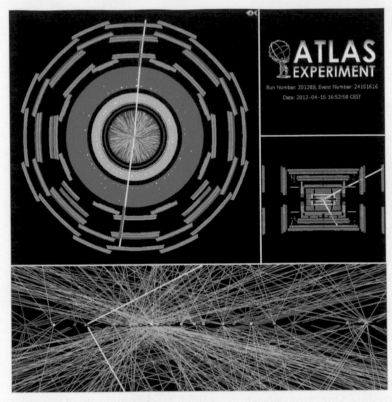

Figure 3.15 An event captured by the ATLAS detector containing two energetic tracks emerging from the same collision point. These are clearly visible in spite of the presence of 25 other low-energy collisions piling up on top of the main collision. These 25 collision points are crammed into a space less than 3 inches (7.8 cm) across.

Source: ATLAS.

The third tracking device is the most voluminous, but also the least accurate. We call it the TRT, an acronym for Transition Radiation Tracker. This detector is made of small carbon fiber straws containing a gas that releases electrons during the passage of a charged particle. Every straw has a fine electric wire positioned in its center to collect these electrons and cause a small electric discharge when hit. On average, a charged particle passes through 32 straws, leaving that many dots in this detector. Thanks to polypropylene fibers placed between the straws, this detector can not only reconstruct the trajectories of particles but also help to distinguish electrons from pions using the electromagnetic radiation they emit when they cross from the fibers to the gas and vice versa. This radiation is more important for electrons than for pions, and so one can distinguish between them. Knowing the exact identity of every particle (or fragment) is essential to allowing us to better reconstruct an event and determine what particle was initially produced.

The electronic modules connected to these three subdetectors carry signals from them, supplying a list of all the channels that were hit. Finally, for each charged particle, we obtain three dots from the Pixel Detector (four after 2015), as seen in the gray circle around the center in Figure 3.16, four from the SCT (black circle), and on average 32 dots from the TRT (the outer zone, in purple). To reconstruct the trajectory of a particle, all we need to do is to connect all these dots (this is clearer in the enlarged view on the right). This event was captured by the ATLAS detector at the beginning of 2009 when the

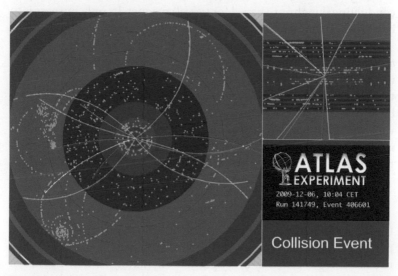

Figure 3.16 The signals left by a charged particle in the three tracking systems of ATLAS. The trajectory of the particle is reconstructed simply by connecting all the dots together.

Source: ATLAS.

LHC operated at lower intensity and fewer collisions were recorded simultaneously.

The magnets

A magnet can bend the trajectory of a particle carrying an electric charge. A positively charged particle will have its trajectory bent in one direction and a negative particle in the other direction, as can be seen in Figure 3.16. The faster the particle, the more difficult it is to bend its trajectory. This is exactly like a car taking a bend at high speed. The faster the car is moving, the more force it takes to make it turn sharply. This force comes from the friction between the tires and the road. It is easy to make a sharp turn at low speed but much more difficult at high speed. This is also why the LHC had to be so big. It would not have been possible to build magnets powerful enough to maintain the protons in an orbit with a sharper curve.

ATLAS has two magnets. The first one, a solenoid magnet, surrounds the tracking systems and bends the trajectories of all charged particles crossing the trackers. But the big pride of ATLAS is its immense superconducting doughnut-shaped (or *toroidal*) magnet. Its only purpose is to bend the trajectories of muons, including the most energetic ones.

A particle that has little speed does not have much energy or much *momentum*—the product of its mass and its speed. Its trajectory can easily be bent so that it will not reach the calorimeters, the layer after the trackers. The solenoid magnet enables us to do a bit of house-cleaning by eliminating the low-energy particles coming from all the other underlying collisions. Since the magnetic force supplied by the magnet is fixed, it is enough to measure the curvature of the trajectory of a particle to determine its momentum.

The calorimeters

Unlike a tracker, a calorimeter (Figure 3.17) must be as massive as possible to stop even the most energetic particles. Its role is to measure the energy carried by every particle emanating from the collision point. Together, the calorimeters are sensitive to all particles except neutrinos. They are of two types: the electromagnetic and hadronic calorimeters. As their names suggest, the first one intercepts any particle that reacts to the electromagnetic force, such as photons and all

Figure 3.17 One of the ATLAS calorimeters during installation.
Source: ATLAS.

charged particles. But photons and electrons are the only particles that will lose all their energy in this device, since the energy loss process very quickly becomes less effective for more massive particles. The second type of calorimeter interacts only with hadrons, the particles made of quarks; these particles may or may not be electrically charged. Protons, neutrons, pions and other hadrons will lose all their energy here.

The muon detectors

The last layer of the detector is dedicated to muons. As you may recall, a muon is similar to an electron but two hundred times heavier. Owing to its mass, it loses very little energy in the electromagnetic calorimeter. And, since it is not made of quarks, it does not interact with the hadronic calorimeter. It is the only charged particle capable of passing through both calorimeters, and hence the only one able to reach the very last layer of the detector, named appropriately the muon detectors (Figure 3.18). These form in fact a tracking system, and provide information to reconstruct muon trajectories.

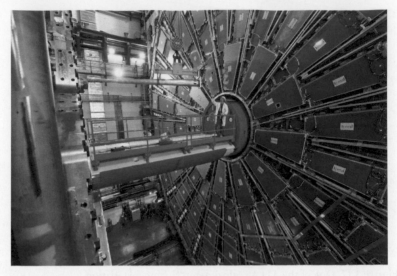

Figure 3.18 It's not easy to get a good selfie in front of one of the two giant Muon Wheels of the ATLAS detector.
Source: CERN.

Particle identification

By combining the information received from the tracking system, the curvature of the trajectory, the energy deposited in the calorimeters and the signals from the muon detectors, we can guess the identity of every particle emerging from a collision, as indicated in Figure 3.19. This works exactly like when one finds tracks in fresh snow. A sufficiently well-informed person can easily distinguish the footprints of a fox from those of a hare or a skier. Likewise, particles leave different imprints when crossing the various layers of the detector. Charged particles, represented by solid lines, leave a signal in the trackers but not the neutral particles, shown with dotted lines. An electron and a photon are easily distinguishable. Both deposit all their energy in the electromagnetic calorimeter, the first calorimeter encountered by a particle moving outward from the collision point, but only the electron leaves a track in the tracking system. A proton is also distinguishable from a neutron since a track is associated with its energy deposit, whereas a neutron only leaves a signal in the hadronic calorimeter. The muon is the easiest particle to

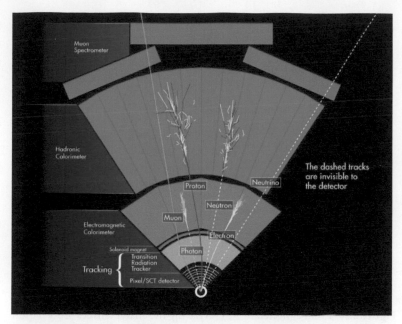

Figure 3.19 Characteristic signatures of various particles in the different layers of the ATLAS detector.

Source: ATLAS.

identify since it leaves a track in the trackers and in the muon detector, with hardly any energy deposit in the calorimeters.

We can even detect the presence of "invisible" particles, those that do not interact with the detector, such as neutrinos for example. These are represented in Figure 3.19 by a white dashed line. Since all events must conform to the principle of energy conservation, each event must be balanced in energy in all directions. Just as with fireworks, we always observe fragments leaving in all directions. One must reconstruct all tracks belonging to the same collision, take into account the energy deposited in the calorimeters and, finally, make sure that everything is balanced, at least in the plane perpendicular to the beams. Before colliding, the protons do not move in this plane and, consequently, neither should the decay products after the collision.

In Figure 3.20, the straight red line in the left diagram represents the track of a very energetic muon. It is easy to see that we are dealing with a muon since this track has left a signal in the muon detector, represented

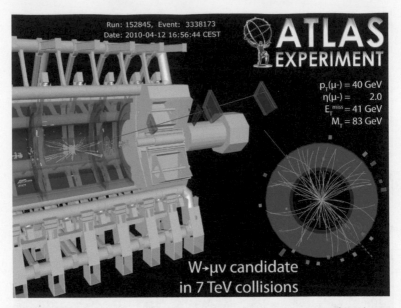

Figure 3.20 Event collected by ATLAS having the characteristics of a W boson decaying into a muon (red line) and a muon neutrino (dotted red line).
Source: ATLAS.

by the panels in green that it is passing through. The right diagram gives the projection of all the tracks reconstructed for this event in the plane perpendicular to the beams. The muon track is indicated by the straight red line. The other, curved tracks (in orange) indicate the presence of particles of low energy coming from the other collisions that occurred simultaneously. The dotted line indicates the direction of the missing energy estimated after adding together all the energy deposited in this event. We associate this missing energy with a particle that has not interacted with the detector, something invisible to it, which has escaped from it taking away some of the energy of the event. In this case, the missing energy is attributed to a muon neutrino. By combining the energy of the muon and the neutrino, we obtain a particle with a mass of 83 GeV, which is more or less the mass of a W boson. This event most likely corresponds to a W boson decaying into a muon and a muon neutrino, although it could also result from another type of event imitating its signature. In particle physics, everything is a question of statistics and one can never be 100% sure that an event is of a precise type, since there is always some background noise, as we will discuss in the next chapter.

The Trigger

An event is a reconstructed image of the decay of a particle produced after two protons collide. Every layer of the detector supplies part of the information needed. The ATLAS detector contains 100 million different channels. Each event corresponds to the reconstructed image of a 100-million-piece jigsaw puzzle. We reconstruct the initial image from the 100 million small pieces of information produced by all the subdetectors. This is exactly what a 100 megapixel camera does, where each image is recomposed from 100 million small dots, or pixels. There is a big difference, though: the ATLAS detector takes a snapshot of every event, and there are up to 40 million events per second. But there is no way that we can save that much data, so many of these snapshots will be thrown away. Still hard to beat!

The detector acts like a tourist on holiday, taking pictures almost continuously. In fact, it takes a snapshot every twenty five billionth of a second, the time elapsed between the passage of two proton bunches. This yields the mind-boggling rate of 40 million pictures (or events) per second. Without a grain of common sense, we would soon be overwhelmed with data, just as we are stuck with piles of photos to be sorted when we return from holiday. We must therefore decide in advance what kind of events are worth keeping. This is the role of a complex system called the Trigger.

For the ATLAS experiment, this selective sorting is done at two levels. In the first stage, ultrafast electronic modules determine in about 2 millionths of a second whether the event that has just occurred is potentially interesting. They do so by looking for signals caused by particles hitting specific parts of the detector. For example, an energetic muon detected in the muon system could have come from a potentially interesting heavy particle produced in a very energetic collision. So could a high-energy deposit in the calorimeters. At this stage, we retain *only* 75,000 events per second.

Then, huge computer networks take over to estimate the potential of each of these events in more detail. A series of rather simple algorithms that can be executed very quickly select the most promising two hundred events every second by looking for striking features. Only these will be kept. The others go directly to the trash can and are lost forever. There is no second chance. We must get it right, even if at this stage we do not have time to reconstruct the events in all their details.

The saved events are then distributed all over the world, using a huge computer network called the Grid, for their final and full reconstruction. In the case of ATLAS, this network distributes the task to hundreds of thousands of interconnected computers located in 11 different countries. Once reconstructed, these events are handed over to the physicists to be sorted and scrutinized from every possible angle.

That is the data analysis phase, where physicists look for new particles. We shall see how this works in the next chapter.

THE MAIN TAKE-HOME MESSAGE

Everything is gigantic when we want to explore the world of the infinitesimally small. The two main tools used in particle physics are accelerators and detectors. An accelerator, such as the Large Hadron Collider, accelerates protons to bring them into head-on collisions at near the speed of light. New particles materialize from the energy released during these collisions. Four big detectors are located around the 17 mile (27 km) LHC ring to detect the fragments of the newly produced particles when they break apart. The detectors are made of concentric layers, each one extracting part of the information that is needed to reconstruct the particles produced in the proton collisions. A detector thus acts like a huge camera taking snapshots of these events, which are rebuilt like a jigsaw puzzle from one hundred million small pieces of information. The last task consists of sorting all these events to extract the most interesting ones, those revealing the existence of new particles or phenomena.

4

The Discovery of the Higgs Boson

We had a great accelerator, the Large Hadron Collider, and state-of-the-art detectors. We switched everything on and it worked *almost*[1] right away. We proceeded to collect billions of events. Then what? How do we find a Higgs boson (Figure 4.1)? Here is how the long process of data analysis goes.

The Higgs boson is a highly unstable particle that survives only a mere 10^{-22} seconds after its production, that is, one ten thousandth of a billionth of a billionth of a second (in other words, not very long). It breaks apart almost immediately, producing other particles. This does not mean that the other particles were contained in the Higgs boson, but simply that the energy equivalent to the mass of the Higgs boson reappears in the form of smaller particles. Hence, we never observe the Higgs boson itself but only its decay products. For a particle, decaying is like making change for a big coin. For example, a one-euro coin can be exchanged for various combinations of 10, 20 or 50 cent coins. Likewise, all particles such as the Higgs boson can give change in multiple ways. Each separate way is called a *decay channel*.

Decay channels

Using the Standard Model, theorists can predict the probability of observing each decay channel (the number of times a particle breaks apart in a particular way), but these predictions depend on the exact mass of the Higgs boson. However, we did not know its mass before its discovery. This is a bit like trying to get very important radio messages without knowing the frequency of the radio station. This is not easy,

---

[1] Everything worked marvelously well right from the start on September 10, 2008, but a major incident occurred nine days later. It caused considerable damage, stopping the accelerator for more than a year.

Figure 4.1 If only finding a needle in a haystack was all it took, discovering the Higgs boson would not have been so hard. But how should one proceed when there is enough hay to fill countless barns?

Source: Marion Hamm.

especially if the signal is weak and there is a lot of "static" noise. For the Higgs boson, when the ATLAS and CMS detectors began recording events, we did not know its "radio frequency": we only knew that its mass had to be over 114 GeV and below 157 GeV (the mass of a proton is close to 1 GeV), since other experiments conducted prior to the LHC had found nothing outside these values.

As we saw in Chapter 2, the theory postulates that the mass of a particle depends on the strength of its interaction with the Brout–Englert–Higgs field. Heavy particles interact more with the Higgs boson. In other words, Higgs bosons prefer decaying into heavy particles. Assuming that these predictions of the Standard Model are right gives us a good idea about the possible ways a Higgs boson will decay, even without knowing its exact mass. The heaviest particle is the *top* quark, with a mass of 174 GeV. But a Higgs boson with a mass between 114 and 157 GeV would have a hard time producing one top quark and one top antiquark, given how heavy they are. A better option is then to decay into a pair of *b* quark and *b* antiquark, the next heaviest quarks after *top* quarks.

Decay into quarks

Unfortunately, there are other ways to produce pairs of *b* and anti-*b* quarks, making it difficult to distinguish them from those coming from the decay of a Higgs boson. Furthermore, as soon as there are quarks in the decay products, it is harder to see clearly what is happening, since quarks never come alone. They always surround themselves with other quarks, and these form hadrons (the class of particles made of quarks).

Quarks are generally produced in pairs and are connected together by gluons that act like rubber bands. Imagine that the ends of the rubber band represent the quarks. When two quarks try to get away from each other at high speed, the rubber band eventually breaks. One ends up with two small segments of the rubber band, each one having two ends. In our analogy, we would then have four quarks. As they are also produced with a lot of energy, these quarks too will continue moving apart, until they break their own rubber band, producing yet more quarks. In turn, all these quarks will form new light hadrons.

In the end, we obtain *jets* containing several hadrons. As there are many particles in a jet, their energy is more difficult to measure than that of individual particles such as electrons, photons or muons. It is therefore not so easy to obtain high-precision results with these jets of particles. This is why the decay channels containing quarks were not used to discover the Higgs boson, even though many of the Higgs bosons produced decayed to *b* quarks.

Choosing between abundance and cleanliness

Some decay channels are more abundant and others are cleaner, that is, more free of background. Unfortunately, these two qualities rarely come together. Although at first glance it would appear wise to look for Higgs bosons in the most frequent decay channel, this is not always the best approach, owing to background.

Let's go back to our example of a radio station of unknown frequency. It is better to try to detect its signal with a device that filters out the noise better than to use an ultrasensitive receiver with which we will hear mostly background noise. Nevertheless, scientists must examine all possibilities to get the overall picture. Hence, we tried to measure the mass of the Higgs boson in several different channels to make sure everything was coherent. This way, we could also check whether or not nature behaves according to the theoretical predictions.

Signal and background

The Standard Model predicts that a Higgs boson can sometimes decay into two Z bosons. But the model also predicts that it is much easier to produce two Z bosons directly, without involving any Higgs boson. So if we find two Z bosons in one event, this does not necessarily indicate the presence of a Higgs boson. It is much more likely, in fact, that these two Z bosons come from other, more run-of-the-mill processes. These well-known processes get in the way when we are looking for something rarer, such as a Higgs boson decaying into two Z bosons.

There are thus two categories of events containing two Z bosons: the *signal* denotes all events containing a Higgs boson, while the *background* refers to all other sources of events. In our radio station example, the signal is the radio message and the background is the static noise. If there is too much static and if the signal is too weak, there is no way we will manage to distinguish the signal from the background. We won't hear anything.

Decays to four leptons

Just like Higgs bosons, Z bosons are unstable and short-lived. One way they can decay is by producing a pair of leptons (that is, two muons or two electrons), although they decay into quarks ten times more often. But once again, there are so many background events containing quarks that it is almost impossible to find such a signal. It would be like trying to locate a cricket by ear during a heavy metal concert! In the end, it is simpler to select events that are less frequent but easier to identify, such as those containing four muons, four electrons, or two muons and two electrons. Certainly, there are fewer of these events but, on the other hand, there is much less background. We can thus find our elusive radio station more easily.

To select the signal, we must apply selection criteria to retain only the events containing two Z bosons. The combined energy of each muon or electron pair has to correspond to the mass of a Z boson or something close to it. Let's go back to the analogy between the decay of a particle and making change for a coin. If the coins in our hand all came from making change for a single coin, their sum would always give the value of the initial coin. But if the change came simply from emptying our pockets, we would have random amounts, since these coins would not originate from a unique coin of given value.

The same thing happens for two electrons or two muons coming from sources other than the decay of a Z boson. Their combined mass will give random values. We can therefore reject all events where the combined mass of the two electrons or muons is not compatible with that of a Z boson. After we have selected all events containing two Z bosons, the last remaining task is to determine which ones come from a Higgs boson. Same scenario: we combine the mass and energy of the two Z bosons and see if they all correspond to the same value. All the events coming from a Higgs boson will end up with the same mass value (with some leeway, as explained in the box "The mass: not a unique value"), whereas the combinations corresponding to the background will give a wide range of random values.

The mass: not a unique value

Just to complicate things a bit more, the mass of a particle does not always have the same identical value like the value of a coin. In particle physics, there is some vagueness in the exact mass of a particle. When decaying, particles give change a bit like in Canada, where it is rounded off to the nearest 5 cents.

This uncertainty in the mass is called the *width* of the particle. Z bosons do not all have exactly the same mass. And, as is often the case in particle physics, everything is a question of probability. The curve in Figure 4.2 gives the probability of measuring a given mass for a particle. The most likely value is the central value, for example 91 GeV for a Z boson. A curve such as that in Figure 4.2 would be obtained if we were to measure the mass of a number of particles, for instance a few hundreds or thousands of Z bosons.

At mid height, as indicated by the horizontal line on the graph, the distance between the two points on the probability curve gives the width. This width is related to the lifetime of the particle. In mathematical terms, one calculates the lifetime of a particle by taking the inverse of its width. The more choice a particle has in terms of decay channels, the more quickly it can break apart. Its lifetime decreases and its width increases.

continued

The mass: not a unique value (*continued*)
....

This is a bit like when several airline companies are offering flights to the same destination: it is then quite easy to find a seat. In contrast, if the choice of airlines is limited, it will be harder to find a seat. Likewise, if a particle has difficulty in finding a decay channel, it takes more time to decay. Its lifetime gets longer.

The Standard Model predicts that the width of the Higgs boson is 4 MeV. This is much less than the widths of the Z and W bosons, which are 2500 and 2000 MeV, respectively, corresponding to approximately 2.5% of their mass. The "natural" peak of the Higgs boson, centered around 125 GeV, is thus much finer than that of the Z and W bosons. When we measure the width experimentally, the resolution of the detector adds to this natural width. The difference in width confers on the Higgs boson a lifetime almost 500 times longer than that of the Z and W bosons.

We can reconstruct the mass of a Higgs boson from events containing two Z bosons. If we measure the mass of several Higgs bosons in many different events, the distribution obtained from the various mass values will give a curve similar to the one in Figure 4.2. If, on the other hand,

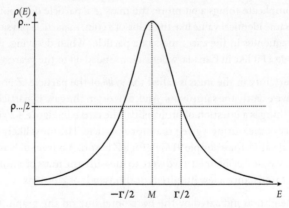

Figure 4.2 The mass of a fundamental particle is not fixed but varies: it can take different values. The probability of measuring a particular value is given by the vertical axis of this graph. The most likely value is the central value, M. This is the value that is given when we speak about the mass of a particle.

Source: Wikimedia.

The mass: not a unique value (*continued*)

the two Z bosons come from the background, we obtain random mass values instead of a peak.

The mass of a Z boson is 91 GeV and that of a Higgs boson, as we shall find out soon, is 125 GeV. In principle, the Higgs boson is too light to produce two Z bosons, since each one has a mass of 91 GeV. But the masses do not add up simply, like taking 91 GeV plus 91 GeV, since the value of the mass can vary. It is like the situation with money and buying power. In principle, with only $125 in your pocket, you cannot buy two articles costing $91 each. This is true unless you find one of them on sale at a reduced price somewhere, which is more difficult but not impossible. A Higgs boson can afford two Z bosons only if one of them is "at a reduced price," that is, far from its central mass value. However, the probability of finding a particle with a reduced mass gets smaller the further away you move from its central value. Consequently, it is rather rare to see a Higgs boson with a mass of 125 GeV giving two Z bosons. This decay channel thus suffers from not occurring often but, on the other hand, the background level is manageable.

Another channel that could a priori be a goldmine is the decay of a Higgs boson into two W bosons, since the W boson is lighter than the Z boson, with a mass of 80 GeV rather than 91 GeV. The decay channel into W bosons is more frequent, even though the Higgs boson prefers decaying into heavy particles, since it can afford it more often. One W boson can decay into a pair of quarks but, as mentioned earlier, there is always too much background. Otherwise, a W boson can split into a muon and a neutrino, or into an electron and a neutrino. As neutrinos are not detected, this adds more complication and the method is less precise, since we have to estimate the neutrino energy from the missing energy. This channel was thus used only to verify the existence of the Higgs boson and not to estimate its mass, at least not at the time of its discovery, when less data were available.

More complex decay channels

The Higgs boson can also decay indirectly via other particles, as is the case with decays into two photons. As photons have no mass, they do not interact directly with the Higgs boson. However, the Higgs boson can decay indirectly into photons by going through an intermediary

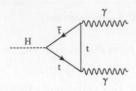

Figure 4.3 Diagram illustrating how a Higgs boson (denoted "*H*") can decay into two photons (γ) through virtual *top* quarks (*t*).
Source: Ulrik Egede

inside "loop" involving *virtual* particles as in Figure 4.3. Virtual particles are produced ephemerally by "borrowing" the energy necessary for their formation for a fraction of a second. That's like buying using credit. But, just as with banks, borrowing is never simple, and this kind of decay occurs only very rarely.

The Higgs boson, represented by the letter *H* in Figure 4.3, splits into two virtual *top* quarks, denoted by *t*. These are not real *top* quarks, as their mass of 174 GeV is far too heavy to be produced from a Higgs boson of mass 125 GeV. The third *top* quark interacts with the first two to give two photons. In the end, the three *top* quarks intervene virtually, leaving only the two photons, represented by the Greek letter γ (gamma).

We can also obtain two photons by inserting other heavy virtual particles into loops or through other even more convoluted processes. Given their complexity, these processes are extremely rare although still possible. All have the characteristic that when the energy of the two photons is combined, it gives the Higgs boson mass. At first sight, this may not seem to be the best way to discover the Higgs boson. Nevertheless, this channel played an essential role in its discovery, as we shall see.

Simulations of events and calibration

We cannot determine how many events come from the signal if we do not know exactly how many emanate from the background. This is why physicists turn to an essential tool: the simulation of events. These simulated events are exact replicas of the signals left by particles in the detector when two protons collide. These events are a concentration of all the knowledge that has been gathered from all particle physics experiments ever conducted

Simulations of events and calibration (*continued*)

over the last few decades. Theorists synthesize all this acquired knowledge and turn it into a prediction of the probability of producing different particles, as well as the odds that these particles will decay through a particular decay channel.

Then the experimentalists simulate the passage of all the particles emanating from these decays through their detector to artificially produce simulated events that replicate real events as faithfully as possible. Achieving a nearly perfect duplication of reality requires considerable effort, since every conceivable aspect must be checked. We have to simulate two things: all known physical processes and the response of millions of detector channels to the passage of diverse types of particles. In addition, we must take into account the fact that we never observe a single collision, but on average about twenty to forty low-energy collisions occurring simultaneously. These simulations bear the name of *Monte Carlo simulations*, in reference to the casinos in Monte Carlo since in particle physics, nearly everything boils down to a question of probability.

Calibration

Before going further, and even before being able to use the simulations, we must make sure that the hundred million detector channels are properly calibrated. To do so, we constantly measure and remeasure a multitude of well-known quantities to cross-check the calibration of the whole detector. This is a daunting task. We must make sure, over and over again, that the energy and position of the particles are precisely measured and do not depend on external factors. For that purpose, we keep track of the ambient humidity, the changes in atmospheric pressure, the failures of certain components, the composition of the various gases used by diverse detectors, the temperature in every nook and cranny of the detector, and many other variables.

When all detector layers are perfectly calibrated, the data will reproduce exactly all of the well-known quantities (Figure 4.4). The next step is to compare the simulations with real data, to calibrate the simulations. This process is constantly evolving. We compare hundreds of quantities measured by the detector with the same quantities in the simulated events. Only then can we be sure that the selection criteria applied to the simulations

continued

Simulations of events and calibration (*continued*)

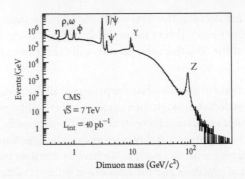

Figure 4.4 The horizontal axis of this graph shows the combined mass of muon pairs identified by the CMS detector at the very beginning of operation in 2010. The peaks correspond to the masses of various particles (ρ, ω, J/ψ, Y(1S), Y(2S) and the Z boson). The vertical axis gives the number of muon pairs found for each mass value. As these particles have been identified numerous times by other experiments in the past, their mass values are very well known. Comparing the masses measured using the CMS detector with the known values allows us to calibrate the detector. Note that both axes use logarithmic scales.

Source: CMS.

will have exactly the same effect on the real data. The last stage is to ensure that all the selection criteria applied to the simulations and to the data are identical, from the collision conditions and trigger algorithms to the behavior of the particles.

To find Higgs bosons, one must first find two Z bosons and then recombine them. For Z bosons, we can select all events containing two energetic electrons or muons and then combine them to obtain the distribution of the combined mass values for all these pairs of leptons. In principle, we should find the mass curve of a Z boson with a central peak corresponding exactly to the mass of the Z boson, stacked on top of various backgrounds. We can verify that the simulated events and the real events give exactly the same curve. Otherwise, we must find which parameters of the simulation code must be adjusted.

The slightest modification to one of the numerous parameters of these simulations can impact negatively on other variables. It is really a delicate exercise, similar to the construction of a house of cards. As soon as a team suggests modifying one parameter of the simulations to improve

> **Simulations of events and calibration** (*continued*)
>
> the agreement with real data for one subdetector, all other groups have to estimate the impact the proposed modifications will have on the other parts of the detector or another physical process. We must make sure that it actually improves the quality of the agreement between the simulations and the real data in every possible respect.

How not to bias the results

Simulations (see the box "Simulations of events and calibration"), which are based on theoretical knowledge and replicate the functioning of the detector, are used to predict the expected results. They are essential tools to avoid biasing the measurements. All of the selection criteria considered in order to discover new particles, such as the Higgs boson, must be established strictly from simulated events. Any transgression of this rule is forbidden, since it can bias the results. Up until the very last minute, physicists doing data analysis will look at real data only to check the calibration and the agreement between the data and simulations, and never to establish the search strategy.

The simulations reproduce not only all the well-known background processes, such as the production of two Z bosons, but also the signal, such as a Higgs boson decaying into two Z bosons. All hypotheses considered by the theorists, even the most far-fetched, are simulated and compared with experimental data in the hope of uncovering new phenomena.

To design an analysis and find the Higgs boson, we first examine the characteristics of the signal and of the background using the simulations. We can then establish the best selection criteria that will eliminate the most background while retaining as much signal as possible. Once these criteria have been established, they become unchangeable. It is thus necessary to make sure that the chosen selection yields the best signal-to-background ratio.

Statistical methods

If I have four 50 cent coins, who could tell me whether these coins came from making change from a two-euro coin or two one-euro

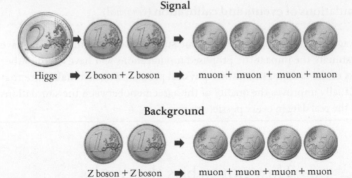

Figure 4.5 The same small change can come from making change for various coins, just as certain events (the background) can imitate the signature of the type of events that we are looking for (the signal).
Source: Pauline Gagnon, Pixabay.

coins (Figure 4.5)? When these coins are particles, we can almost achieve that, but we must rely on very sophisticated, advanced statistical methods.

For the discovery of the Higgs boson, the approach used consisted in estimating how many events coming from the background could get through the selection criteria, designed to retain a maximum of signal. This can be done in two ways: either by using the simulations or by evaluating the background directly from the data. For example, if we are looking for Higgs bosons in a particular mass region, we can evaluate the background level in another mass region and then extrapolate to the region of interest.

Once the selection criteria are set in concrete, we apply them to the real data. We can then observe whether the retained events correspond to the simulations of the background only or whether there is a small excess imputable to our signal.

Recipe for boson syrup

We often hear the expression "Higgs hunting" in relation to the search for the Higgs boson, as if we were going to find one, shoot it, stuff it and hang it on the wall. Nothing could be more remote from

Figure 4.6 The Lacoudès Sugar Shack, near Sainte-Rose-du-Nord on the Saguenay River in Quebec, Canada. Looking for a Higgs boson is very much like making maple syrup.

Source: Yves Lagacé.

reality than that. It is not about hunting but, rather, much more like gathering. In fact, it is very much like producing maple syrup (Figure 4.6). So here is my recipe for Higgs boson syrup.

To make good maple syrup, one must first find the right type of trees, namely sugar maple, and avoid tapping other species, such as birch or ash trees. The sugar in sugar maple sap is our signal and the water in it is our background. Collecting sap from other trees, especially other maple species that look alike but whose sap is much less sweet, only adds to the background. It dilutes our signal. One must collect the maple sap drop by drop, just as we accumulate events over time, one by one, every time protons collide in the LHC. To extract 1 liter of maple syrup, one must boil off 27 liters of maple sap. Likewise, one must collect five billion events in the hope of finding one single Higgs boson.

Error margins

One must always take error margins in experimental measurements into account. The numbers of signal and background events can fluctuate, since particle physics does not obey fixed laws but rather statistical laws. Here is an example to illustrate how this works. Imagine a bag filled with marbles where half of the marbles are green and the other half blue. Suppose I ask you to estimate the percentage of green marbles in the bag by letting you take a sample of ten marbles. How many of these ten marbles will be green? Five? Six? Two? All of these values are possible, although drawing five green marbles is certainly more likely than getting only two.

On the other hand, if you were to draw at random not ten but one hundred marbles, what percentage of green marbles would you obtain? Any value between 45% and 55% is very likely, although slightly more or less is also possible. But if you take a sample of one thousand, ten thousand or even more marbles, the odds are excellent that the percentage of green marbles will be very close to 50%. The larger the measurement sample (the number of marbles drawn at random), the better the chances of finding the real answer, that is, that 50% of the marbles are green. However, when dealing with very small samples, it is not surprising when one obtains values very far from 50%, such as 20% or 30% for example.

When we select events with the aim of extracting a signal, the numbers of both signal and background events can vary considerably, especially if the sample selected is small. These *statistical variations* are taken into account and added to the experimental errors to determine the total *error margin*. This is set large enough that one has a 68% chance of having the right answer within this interval. By definition, this is one *standard deviation*. And there is a 95% chance that the right answer lies within two standard deviations.

When looking for a possible signal, one compares the signal intensity with these fluctuations, after the background has been subtracted. A signal at least five times the size of the possible statistical fluctuations (combining both the signal and the background fluctuations) corresponds to the criterion of five standard deviations (or 5 sigma) used in particle physics. The odds of getting a background fluctuation bigger than five standard deviations are one in 3.5 million.

That is the criterion that gives researchers the right to open a champagne bottle (Figure 4.7).

Figure 4.7 Professors François Englert and Peter Higgs in the midst of a lively discussion just after the announcement of the discovery of the Higgs boson at CERN on July 4, 2012. The two men had never met before.
Source: CERN.

The last stretch

Of course, it would have been easier to wait several more months, accumulate larger samples of events and take extra time to analyze everything calmly. Except that in the summer of 2012, the competition for the discovery of the Higgs boson between the CMS and ATLAS teams was fierce. The biggest conference in particle physics of the year was due to start on July 4. Both groups wanted to present their most recent results there. Furthermore, the credibility of the whole Large Hadron Collider project was at stake. People all over the world were looking forward to seeing if the Higgs boson would actually make its appearance or if its prediction had been pure fantasy. The pressure was at its peak. The members of the ATLAS and CMS Collaborations made enormous efforts, in terms of both ingenuity and efficiency, to extract the maximum information from the data available at the time. These efforts were highly rewarded, since the bar of five standard deviations was reached by both experiments.

The discovery

Both teams accumulated data until the week preceding the conference to maximize the size of their data sample. Time was needed to complete the mandatory calibration steps and quality checks on the latest recorded data. Despite having teams of people working in various time zones, many key individuals were working day and night since there was absolutely no time to spare. A few days before the conference, the selection criteria established using the simulations were applied for the first time to the real data to reveal, at long last, how many events passed the selection criteria (Figure 4.8). Here is what the ATLAS Collaboration presented on July 4. The results of the CMS experiment were just as impressive.

In Figure 4.9, the vertical axis gives the number of events collected, while the horizontal axis displays the combined mass measured in GeV for the four leptons (muons or electrons) for all the events satisfying the selection criteria. The part in red corresponds to events coming from the main background, two Z bosons produced directly, as established

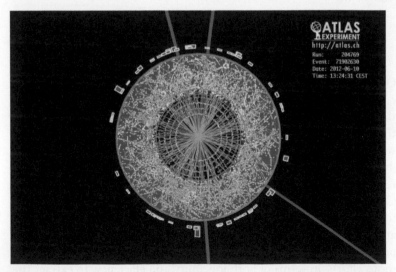

Figure 4.8 An event captured by ATLAS having the characteristics of a Higgs boson decaying into two Z bosons, each one giving in turn two muons. The four red lines indicate the muon tracks.

Source: ATLAS.

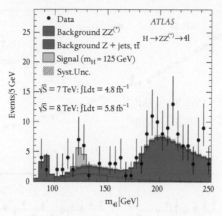

Figure 4.9 One of the diagrams shown on July 4, 2012 by the ATLAS Collaboration to prove the discovery of a new boson. The vertical axis gives the number of events found, all of them satisfying the criteria designed to select events containing a Higgs boson decaying into four leptons (muons or electrons) via two Z bosons. The horizontal axis gives the combined mass for these four leptons. The simulated background is indicated in red and purple. This corresponds to other types of events that have the same characteristics as the signal but come from other sources. The excess, in pale blue, corresponds to the theoretical prediction for a Higgs boson having a mass of 125 GeV. The black dots correspond to the real data. One must compare the distribution of these black dots with what the simulation predicts from the background (shown in red) and determine if there is any significant excess coming from a source other than the background. In this diagram, this occurs only around 125 GeV.
Source: ATLAS.

from the simulations. The part in purple represents other sources of background. The black dots give the numbers of events found in data collected by the ATLAS detector. The vertical bar associated with each dot represents the size of the possible statistical fluctuations and the experimental error. The hatched area above corresponds to possible fluctuations in the background.

If the experimental data, given by the black dots, had coincided with the background distribution, taking into account all possible statistical fluctuations, we would have concluded that there was no Higgs boson. The background is roughly what we see for all mass values: the black dots reproduce more or less the red area everywhere, except around 125 GeV. There, an excess of events is clearly visible and cannot be

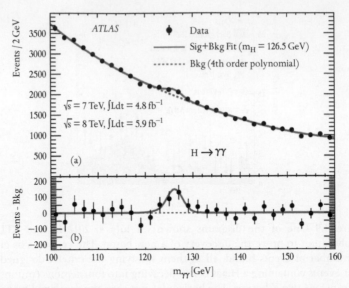

Figure 4.10 The second diagram shown on July 4, 2012 by ATLAS to prove the discovery of the Higgs boson. In this case, the events satisfy the selection criteria corresponding to the decays of a Higgs boson into two photons. The vertical axis gives the number of events found as a function of the combined mass of the two photons. The background (solid line outside the region between 120 and 130 GeV in the top image) corresponds to all events found with randomly produced pairs of photons. The lower plot shows the excess imputable to the new boson after subtracting the background.

Source: ATLAS.

explained in terms of statistical fluctuations in the background in that region. On the other hand, this excess is perfectly compatible with the number of events predicted for the signal by the simulation, namely for a Higgs boson having a mass of 125 GeV, as shown in pale blue.

All this in itself was extremely promising, but before claiming victory, we had to verify that we had also obtained a signal in other decay channels. The most convincing was in the decay channel of Higgs bosons into two photons. Figure 4.10 shows the distribution of the combined mass of the two photons for all of the events selected for this analysis. The black dots represent once again the experimental data and the dashed curve shows the estimated number of events coming from the background. This curve regroups all events containing two independently produced photons.

The amount of background in the region indicated by the dashed line, between approximately 120 and 130 GeV, was extrapolated from the background level observed outside this zone (the solid curve). This method relies entirely on data, without involving simulations. This limits the error margin. If we subtract the number of estimated background events from the data, we obtain the small excess given at the bottom of the figure. Once again, there are more events than what is attributable to the background. If this excess does not come from the background, there must be another source, namely the much anticipated signal: Higgs bosons decaying to two photons.

On July 4, 2012, two decay channels clearly exhibited a significant excess of events beyond the possible variations in the background. These two excesses showed up at essentially the same mass value. Moreover, this observation was made by both the CMS and the ATLAS Collaborations, confirming, without the shadow of a doubt, the presence of a new particle having all the characteristics of the Higgs boson.

The discovery was announced at CERN with great flourish and broadcast live in Melbourne before an audience of 900 physicists attending the opening of a major particle physics conference. But it took eight more months before the CMS and ATLAS collaborations had enough data to confirm beyond any doubt the identity of this new particle, after measuring several of its properties. We had to make sure that this particle not only looked like a Higgs boson, but also sang, walked and danced like one.

This has now been done. By measuring its spin (angular momentum), we were able to verify that the spin had a value of zero, as predicted by the theory. The Higgs boson is the only fundamental particle that has zero spin. Unlike all other fundamental particles, both fermions and bosons, it thus has no privileged direction in space. For that reason, the Higgs boson is also called the *scalar boson*, to stress that its spin is zero.

But there are several possible versions of the Higgs boson. It still remains to be seen if this is indeed the Higgs boson predicted by the Standard Model, the one imagined back in 1964 by Robert Brout, François Englert and Peter Higgs and shortly afterwards by Tom Kibble, Gerald Guralnik and Carl Hagen. It could also be the lightest of the five Higgs bosons postulated by another theory called supersymmetry, something I shall speak about in detail in Chapter 6. So the end of this story has still not been written. It will take time and much more data, which has started to be collected after the LHC restarted in spring 2015, before we have more clues.

Nobel Prize

The confirmation in March 2013 that the new particle was a type of Higgs boson seems to have convinced the Nobel committee that Robert Brout, François Englert and Peter Higgs were right in 1964. Robert Brout being deceased, only the other two received the Nobel Prize in Physics on October 8, 2013 (Figure 4.11). This prize is never awarded post-humously and can be shared by at most three persons or institutions (although no institution has ever received a Nobel Prize, except for the Nobel Peace Prize).

It is a pity that the committee did not choose to award the prize jointly to these two theorists and to CERN, since without experimental confirmation, a theory is not worth more than the paper it is written on. However, the efforts of the thousands of people involved in this discovery were explicitly acknowledged by the Nobel Prize committee, since the prize was granted for a theory that "had recently been confirmed by the discovery of the predicted fundamental particle by the ATLAS and CMS experiments from the Large Hadron Collider (LHC) at CERN."

Figure 4.11 CERN Director General Rolf Heuer addressing a group of physicists from CMS and ATLAS gathered to hear the announcement of the Nobel Prize in Physics on October 8, 2013.
Source: CERN.

Awarding part of the prize to CERN would have been a great way to stress that nowadays particle physics, as well as many other disciplines, requires the concerted efforts of large multinational teams. No individual, not even any nation, can single-handedly push forward research in this field, as we will see in Chapter 8. Nevertheless, the whole laboratory was jubilant on that day, since every one of us knew that our contributions had been essential.

An unforgettable moment

The announcement of the discovery of a new boson by the CMS and ATLAS collaborations at CERN will remain in the memory of every particle physicist of that time. Everyone will remember where he or she was on that day. The announcement was made at 9 a.m. in the morning on July 4, 2012 at CERN in a packed auditorium (Figure 4.12). Several people had queued up

Figure 4.12 The main auditorium at CERN shortly before the announcement of the discovery of a new boson on July 4, 2012. At the door, the line stretched out for several hundred feet (over 100 meters), crossing the main building and the cafeteria and ending outside. Some people had even spent the night waiting at the auditorium door to secure a seat.
Source: CERN.

continued

An unforgettable moment (*continued*)

all night hoping to get a seat, even though the event was broadcast live in several other auditoria at CERN (just as crowded) as well as on the Internet.

The date was deliberately chosen to coincide with the opening of the biggest particle physics conference of the year, held in Melbourne, Australia. That's where I was. But nobody, not even the Director General of CERN, knew much in advance exactly what the two experiments would present until moments before the public announcement. For example, the ATLAS final results were distributed to the collaboration members less than three days before the seminar. The physicists directly involved in these analyses worked day and night to finalize the results in time, taking advantage of different time zones, since they were working in several continents. For sure, most of them had had little sleep over the last few days. The whole community expected interesting results since in December 2011, during a joint CMS–ATLAS seminar, telltale signs were already visible in both groups' data.

Shortly after my arrival in Melbourne on Monday July 2, I entered a McDonald's restaurant to take advantage of their Internet connection. This is where I first saw the latest ATLAS results, those revealing the discovery of the new boson. It was both extremely exciting but also terribly frustrating since there was nobody around with whom I could share my joy. At any rate, we were not allowed to reveal these results before the conference. The whole collaboration had to have the opportunity to read and comment on them first before they were made public. Having no access to the CMS results, I was frantically asking myself the same question I am sure every member of both collaborations was asking him or herself: was the other team seeing the same effect?

This was a legitimate question, since both groups were working totally independently and in complete secrecy. There were of course some rumors but in the end, very little information leaked out in advance to spoil the surprise during the official announcement.

So I settled down in the front row in the Melbourne auditorium where the CERN seminar was about to be broadcast live to the 900 conference participants. My role consisted in commenting on the presentations in a live blog written on behalf of CERN for the Quantum Diaries website, both in French and in English. (For history's sake, here is the original blog I wrote live during the seminar, typos included! http://www.quantumdiaries.org/2012/07/04/live-blog-on-cern-higgs-seminar-from-melbourne/).

An unforgettable moment (*continued*)

Figure 4.13 Part of the audience in Melbourne watching the live broadcast of the CERN seminar where the discovery of a new boson was announced on July 4, 2012. I am sitting in the front row.

Source: Laura Vanags, ARC CoEPP.

To my right, just outside the photo in Figure 4.13, a journalist working for one of the main news agencies was struggling to follow the presentations. I blogged "from one hand" in two languages and listened to the presentations with the other one while distilling the information as it unfolded to assist this journalist. I was operating on pure adrenalin.

The atmosphere was feverish, in great contrast to the usual subdued ambience typical of such meetings. At 8:56, François Englert and Peter Higgs, the first theorists to have proposed the existence of the Higgs boson, walked into the CERN auditorium, triggering thunderous applause in Geneva, as well as in all other auditoria, as though they could hear us. It was the first time these two men had met. At 9:00 sharp Geneva time, 17:00 in Melbourne, a dead silence befell every auditorium. Joe Incandela, spokesperson of the CMS Collaboration, presented CMS results (Figure 4.14). At 9:40, it became obvious that CMS had irrefutable proof of the discovery of a new boson. The audience, which had been holding its breath until then, burst into applause in all of the auditoria.

Then at 10:00, right on the dot, Fabiola Gianotti, spokesperson of ATLAS, took her turn to present her collaboration's results. Just as Joe had done,

continued

An unforgettable moment (*continued*)

Figure 4.14 Joe Incandela, spokesperson of the CMS Collaboration, stands in front of a crammed auditorium on July 4, 2012 to present CMS results at the seminar where CERN announced the discovery of the Higgs boson. This seminar was broadcast live to several other equally packed auditoria at CERN, as well as in Melbourne, Australia, where 900 physicists were attending a major conference.

Source: CERN.

she showed all the careful checks that had been carried out to support the methods used, and finally revealed the results at 10:40: another clear and unequivocal signal. Shouts and applause erupted from everywhere. Fabiola, rather tense and concentrated until then, relaxed at last and started laughing with the audience. Peter Higgs, all teary-eyed, and François Englert, jubilant, delivered their first impressions followed by long, warm applause (Figure 4.15)

Everybody was exultant, even in Melbourne, although we felt slightly remote; but the kangaroos were happily hopping all over. The following reception was lively, and nobody left feeling thirsty. I rushed back to my hotel to draft another blog summarizing the results. That's when the telephone started ringing. Several media organizations from Canada wanted to know the details and, like many of my colleagues, I gave numerous interviews in the following days, both early in the morning and late in the evening, given the time difference. "You finally got promoted from *searcher* to *finder!*" summarized a good friend of mine.

An unforgettable moment (*continued*)

Figure 4.15 The ambience was festive and animated at the press conference following the announcement of the discovery of the Higgs boson in July 2012. Journalists from around the world gathered around Peter Higgs and François Englert to get an interview. A few minutes earlier, the physicist Sau-Lan Wu had stopped Peter Higgs to tell him, laughing: "I have been looking for you for more than twenty years!" His reply to her: "Well, now you have found me."

Source: CERN.

THE MAIN TAKE-HOME MESSAGE

Finding the Higgs boson is similar to looking for a signal from a radio station broadcasting on an unknown frequency. The more static background noise there is, the more difficult it is to find the signal. The signature of a Higgs boson is not unique and can be imitated by other particle decays that have nothing to do with the Higgs boson. A particle decay is similar to making change for a big coin. But, four 50 cent coins can come from giving change for either a two-euro coin (by analogy, the signal) or two one-euro coins (the background). Only advanced statistical methods can allow us to distinguish the signal from the background.

continued

THE MAIN TAKE-HOME MESSAGE (*continued*)

Physicists use simulations to produce made-up events that help us to determine what distinguishes the signal from the background, and to establish selection criteria. Not only do these simulations allow us to understand the detector thoroughly and calibrate it, but they also yield an estimate of how many events will come from the background when we apply our selection criteria to the data. If more events are found than what is predicted to come from all other known processes, the chances are good that we have found a new particle.

Moreover, if such an excess occurs in several different decay channels and is observed by two completely separate experiments working independently and in complete secrecy, the evidence becomes compelling. It is as if several independent teams had determined the same frequency for the mysterious radio station, without consulting each other and using different instruments. This is what was revealed on July 4, 2012 with the announcement of identical results by the ATLAS and CMS experiments. We knew then that we had discovered a new particle that looked just like a Higgs boson.

5

The Dark Side of the Universe

With the discovery of the Higgs boson, one might be tempted to think that we finally have a complete picture of the material world that surrounds us and that all mysteries in particle physics have been solved. Well, that is far from being the case, quite the contrary. In fact, the current theoretical model, the Standard Model described in the first chapter, explains only a mere 5% of the total content of the Universe. Some of you may have already heard about dark matter (Figure 5.1), that mysterious type of matter that is not seen but accounts for 27% of the content of the Universe. Visible matter (you, me and all that we see on Earth, in stars and galaxies) accounts for only 5% of its total content. How do we know that this dark matter really exists? I describe the evidence below.

Before speaking about dark matter, I must say some words about dark energy, since this accounts for 68% of the total content of the Universe. But this part will be brief, since very little is known. In 1998, two independent research teams, one led by Saul Perlmutter and the other by Adam Riess and Brian Schmidt, measured the speeds at which galaxies moved away from each other. Both teams observed not only that the Universe was expanding but also that this expansion was accelerating. This discovery brought them the Nobel Prize in Physics in 2011. As you all know, to accelerate, be it on a bicycle or in a car, requires energy. So where does the phenomenal energy capable of accelerating the expansion of the Universe come from? Nobody knows. Besides, this energy is of a completely unknown nature. It is called *dark energy* to draw a parallel with dark matter. We shall see in the coming sections how scientists from Planck, a satellite-borne experiment launched by the European Space Agency, determined that it amounts to 68% of the content of the Universe.

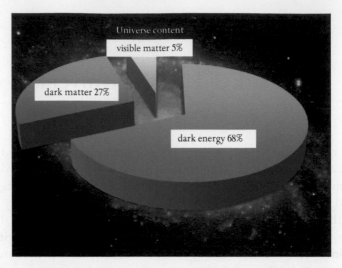

Figure 5.1 Nearly all the content of the Universe consists of unknown substances: a type of matter completely different from what we know, called "dark matter," and a form of energy just as mysterious known as "dark energy." *Source*: Pauline Gagnon.

Dark matter: invisible but omnipresent

The Swiss astronomer Fritz Zwicky was the first person to discover the existence of dark matter, in 1933. He wanted to measure the mass of a galactic cluster (a group of more than one hundred galaxies bound together by gravitational forces) using two different methods. He first estimated the mass from the rotational speed of the galaxies inside the cluster. Just as children playing on a merry-go-round have to hang on to avoid being ejected, galaxies in a rotating galactic cluster need a force to keep them together (Figure 5.2). In this particular case, this force is provided by the gravitational force and is supplied by the matter contained in the galactic cluster. In order to keep everything bound together, there has to be enough matter to generate the necessary gravitational force, otherwise the galaxies would scatter.

Zwicky then verified his calculations by a second method. This time, he estimated the total mass of the galactic cluster from the light emitted by its galaxies. The quantity of light emitted depends on the contents of the galaxy. Therefore, this method yields a rough estimate of the quantity of

Figure 5.2 To prevent the stars of a rotating spiral galaxy from scattering, a force has to keep them in place, just like spinning children have to hang on to the merry-go-round to avoid being ejected.
Source: Nils Brehmer, Pauline Gagnon.

matter contained in a galactic cluster. He noticed that the results did not balance at all. The quantity of matter visible was insufficient by far to produce the gravitational force needed to maintain the cohesion of the galactic cluster. He thus deduced from this observation that a new, unknown type of matter must be generating a gravitational field without emitting any light, hence the name *dark matter* (from the German *dunkle Materie*).

Rotational galaxies

Unfortunately, Zwicky's calculations were imprecise. It was not until the 1970s that the American astronomer Vera Rubin measured the rotational speeds of stars inside a spiral galaxy with enough precision to convince the scientific community. A spiral galaxy is a galaxy that spins at high speed. Rubin observed that the stars in such galaxies were moving at more or less the same speed, no matter how distant they were from the galactic center.

But this contradicts Kepler's law describing the rotation of a star around the center of a galaxy. The further away a star is from its galactic center, the more slowly it should move around it. This is described by curve A in the graph in Figure 5.3. This displays the rotational speed of a

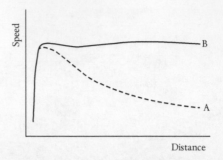

Figure 5.3 The further away a star is from the center of its galaxy, the more slowly it should move around it according to Kepler's law, as indicated by curve A. However, stars from spiral galaxies follow curve B. Their speed is independent of their distance from the galactic center, revealing the presence of an enormous quantity of invisible matter.

Source: Wikipedia.

star according to its distance from the center of its galaxy. But Vera Rubin noticed instead that stars in spiral galaxies followed curve B. It was as if the most distant stars were rotating around galaxies ten times more massive than they were observed to be. This could only occur if enormous quantities of invisible matter filled these galaxies, extending even beyond their most distant visible objects. She was therefore the first person to prove in a more quantitative manner the existence of dark matter. Since then, the evidence has accumulated, as we shall see in this chapter.

Gravitational lenses

The Universe thus contains an incredible amount of matter of an unknown type, called dark matter. Can its presence be detected by some method slightly more tangible than estimating the rotational speed of stars in spiral galaxies? Yes, with *gravitational lenses*, one of the most striking techniques for the detection of dark matter. Gravitational lenses work on the principle that large quantities of matter (either visible matter or dark matter) generate strong gravitational fields. In turn, these fields deform the space around them and modify the trajectories of light. (See the section 'The Brout–Englert–Higgs field' in Chapter 2 for an explanation of the concept of a field.)

Imagine that two people are holding a stretched-out bed sheet and another person tosses a ping-pong ball onto it. The ball will move in a straight line, simply following the surface of the sheet. But suppose that someone has dropped a heavy object such as a billiard ball in the middle of the sheet. The ping-pong ball will then describe a curve to follow the deformed surface of the sheet.

Light behaves just like the ping-pong ball. It has to follow the curvature of the space in which it propagates. Empty space containing no matter is similar to a taut bed sheet. Here, light moves in a straight line. But massive objects, such as stars, galaxies and huge blobs of dark matter, all generate a strong gravitational field. The space around them is deformed and light will follow the curvature of this deformed space. This is what occurs when light passes near the Sun: it is deviated slightly. A person observing light coming from a star placed behind the Sun will have the impression that this light emanates from another, slightly displaced location as illustrated in Figure 5.4.

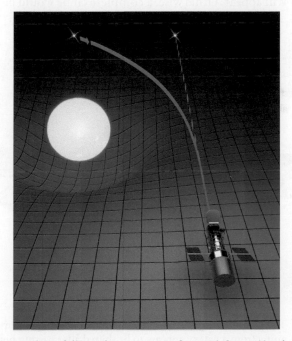

Figure 5.4 An object follows the curvature of space deformed by the presence of a massive body.

Source: David Jarvis.

An accumulation of dark matter acts like a lens. In the illustration in Figure 5.5, two people equipped with a telescope are observing a galaxy located behind a blob of dark matter. This blob is our "lens." Part of the light coming from the galaxy will bend when passing near the blob of dark matter, as indicated in the figure. For the people observing it with the telescope, the galaxy appears to be shifted, as if it was located somewhere else (in the positions of the top and bottom images), since the eye extrapolates in the direction of the incoming light rays. The observers will thus see not a unique image, but several images. Figure 5.5 shows what happens in two dimensions. Figure 5.6 illustrates what occurs in the plane perpendicular to the first plane. In three dimensions, the light is shifted not merely upward and downward as in Figure 5.5, but in all directions.

The light will then form a ring like the one shown in Figure 5.6 and in the photo taken by the Hubble Telescope shown in Figure 5.7. When the galaxy and the telescope are not perfectly aligned, only small arcs will appear. Otherwise, a complete circle would be visible.

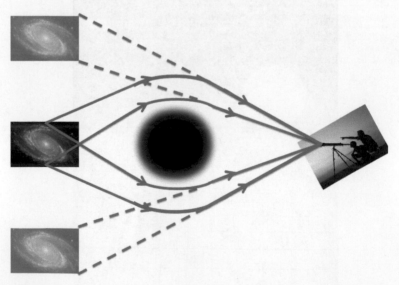

Figure 5.5 The principle of a gravitational lens illustrated in two dimensions. The light coming from a galaxy appears to be shifted after passing near a blob of dark matter. For observers placed on the other side of this dark matter, the light seems to come from shifted positions, above and below the real position.
Source: Pauline Gagnon.

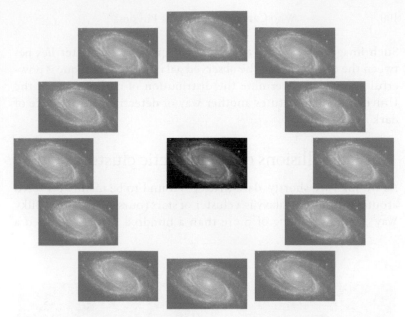

Figure 5.6 In three dimensions, the light diverted by a blob of dark matter forms a ring around the real position of the observed galaxy.

Source: Pauline Gagnon.

Figure 5.7 Dark matter located between a galaxy and the telescope reveals its presence by forming a ring around the image of the galaxy as observed by the Hubble Telescope.

Source: NASA.

Such images reveal that an important quantity of matter lies between the observers and the observed galaxy. This technique is powerful enough to determine the distribution of dark matter in the Universe and constitutes another way of detecting the presence of dark matter.

Collisions of two galactic clusters

As we shall see shortly, dark matter is found to be mainly centered around galaxies. A galaxy is a cluster of stars (ours is called the Milky Way), and a grouping of more than a hundred galaxies is called a

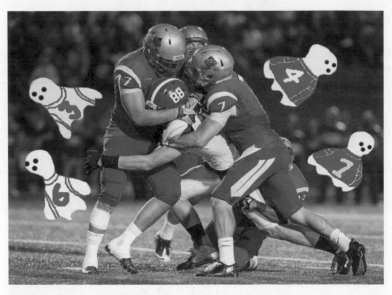

Figure 5.8 Dark matter, just like ghosts, can pass through ordinary matter (represented here by regular football players) without slowing down. A similar situation occurs during a collision of two galactic clusters. The normal, visible matter slows down, whereas dark matter can get through the other galactic cluster and continue without interacting with it. At first, the dark matter and the visible matter overlap in the galactic cluster. After the collision, the two are separated and the dark matter is found to be moving ahead.

Source: Pierre Bonanfant, Pauline Gagnon.

galactic cluster. It sometimes happens that two galactic clusters traveling in opposite directions collide.

To better understand what is happening, try to picture a galactic cluster as a team of American football players. Each player represents a galaxy, and the team forms a galactic cluster that has some cohesion, like a swarm of bees. Imagine also that our football team contains not only normal players but also ghosts (Figure 5.8). The normal players symbolize the visible matter in the galactic cluster, while the ghosts represent the dark matter. We can now simulate a collision of two galactic clusters. It will be similar to two football teams rushing toward each other, each team having both normal and ghost players.

When the two teams collide, the normal players ram into each other and slow down considerably. Eventually, both groups manage to make their way through the pack, warming themselves up in the process from all the resulting friction. But, as everyone knows, ghosts can get through the pack without being slowed down in the least. In the end, the ghosts in each team will find themselves ahead of the pack, having easily overtaken their normal teammates as the latter were slowed down. The collision will have the effect of separating the two types of players, the ghosts taking the lead.

The Hubble Telescope has captured an image of such a collision, known under the name of the Bullet Cluster. It can be seen in the image in Figure 5.9, taken after a collision has occurred between two galactic clusters. The pink area on the left represents the visible matter of the galactic cluster that is moving toward the left, i.e. the cluster that came from the right. The pink area on the right shows the other galactic cluster, traveling toward the right from the left. Under the effects of friction, all this matter has heated up during the collision and generated large quantities of X-rays, which are shown in pink. The areas in purple represent dark matter, whose presence was detected by means of gravitational lenses. The purple coloring was thus added to the picture, whereas the zones in pink correspond to ordinary matter, emitting X-rays. The shift between the dark matter (in purple) and the visible matter (in pink) is clearly visible for both galactic clusters. You can watch an animation depicting this collision at https://www.youtube.com/watch?v=eC5LwjsgI4I.

Figure 5.9 Photo taken by the Hubble Telescope of a collision between two galactic clusters. The zones in purple show where dark matter was detected using gravitational lenses (and has therefore been added to the picture), whereas the pink light comes from X-rays emitted when regular matter overheated due to friction.

Source: NASA et al.

The beginning of the Universe

The Big Bang marks the birth of the Universe. Instants later, it was incredibly hot, with the temperature reaching around 10^{27} degrees (at this temperature, there is no need to specify whether we are speaking about degrees Celsius, Fahrenheit or Kelvin!). It was so hot that only radiation was present. Following an ultrafast cosmic expansion in the first fraction of a second, the Universe continued to stretch out but at a much lesser pace. All the energy it contained was scattered throughout an ever-increasing volume, and the Universe slowly cooled down. The same type of cooling occurs when we release the air contained in an inner tube. It cools as it expands, and one can feel this when deflating a bicycle tire: press on the valve, and you will feel cold air passing over your finger. In a similar way, the temperature of the Universe dropped during the expansion following the Big Bang.

After sufficient cooling, the content of the Universe gradually "materialized" in the form of particles as can be seen in Figure 5.10. At the beginning, quarks and gluons had too much energy and could not bind together.

The beginning of the Universe (*continued*)

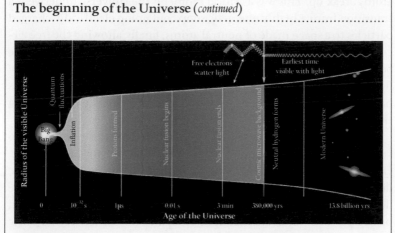

Figure 5.10 The main stages in the formation of matter after the Big Bang.
Source: BICEP2.

Instead, they formed a *quark–gluon plasma* as described in Chapter 3. Approximately 10^{-10} seconds after the Big Bang, the temperature had fallen enough to allow protons and neutrons to form. The Universe still consisted essentially of radiation, and particles of matter constantly appeared and disappeared. Another 380,000 years would be needed before atoms could form, and a billion more years before large structures such as galaxies and galactic clusters appeared.

When the Universe became transparent

In the next three sections, we shall see how astrophysicists from the Planck Collaboration determined the quantity of dark matter in the Universe by studying the cosmic microwave background. Before getting there, we must first talk about the origin of the Universe, namely the Big Bang, a massive explosion that took place 13.8 billion years ago, on a Thursday morning around 7:15 a.m. (see the box "The beginning of the Universe").

The energy released during the Big Bang appeared at first in the form of radiation. Particles started to appear once the Universe had cooled sufficiently under the influence of its expansion. It was not until 380,000 years later, when the temperature had fallen to around 11,000 °F (6000 °C), that atoms started to form, since above this temperature, atoms break up. This was a critical moment: the Universe underwent a transition from a highly energetic soup containing electrically charged particles to a space made of neutral atoms, finally allowing the free circulation of electromagnetic waves such as light. Hence, the Universe became transparent and light was able to propagate freely. Nearly all the light present in the Universe at that time still exists today, since it has had nearly no chance of encountering anything on its passage during the last 13.8 billion years.

How could this be possible? One must understand that the Universe was then and still is today an immense, essentially empty space. Of course, on Earth or in any star, the density is much higher, but the distances between stars and between galaxies are so big that the average density of the Universe amounts to only one proton per 7 cubic feet, or five protons per cubic meter. In comparison, 35 cubic feet (1 cubic meter) of water contains 6×10^{29} protons and neutrons (the two particles have more or less the same mass). If we were to flatten the Universe today into a disk with the density of water (Figure 5.11), it would be reduced to an immense pancake 90 billion light years in diameter (the current size of the Universe) but a mere 3/64 of an inch (1 millimeter) in thickness. It is no wonder, then, that almost all of the light present

Figure 5.11 If all the visible matter contained in the Universe was compressed in one direction until it reached the density of water (62 lb per cubic foot or 1000 kg per cubic meter), it would be reduced to a pancake 90 billion light years in diameter, the radius of the Universe, but only 3/64 of an inch (one millimeter) in thickness. (Based on an idea from John C. Brown).

Source: Pauline Gagnon.

380,000 years after the Big Bang is still wandering around today, having never encountered anything on its passage.

The cosmic microwave background

This fossil radiation, called the *cosmic microwave background*, dates back to the time when the Universe was barely 380,000 years old. If the Universe was a 100-year-old person today, 380,000 years would correspond to when this person was only one day old, in proportion. A baby Universe! This fossil light has been traveling for approximately 13.8 billion years and is reaching us today from all directions.

For electromagnetic waves such as light, there exists a correspondence between the temperature and the radiation that a body emits when heated. The light radiated when the temperature of the Universe was 11,000 °F (6000 °C) corresponded to visible light, like the light produced when we heat a piece of metal until it starts to glow. During the expansion, the energy of the Universe was distributed over a larger volume. It thus cooled, just like when we pour a glass of hot water into a much larger volume of cold water. The drops of hot water give up part of their energy to the whole liquid. Finally, the whole liquid reaches a much lower temperature than that of the initial glass of hot water.

The temperature of the Universe today is no more than −454.8 °F (−270.425 °C), or on the absolute scale, 4.9 °R in degrees Rankine or, in kelvin, 2.725 K. This temperature corresponds to the range of microwave radiation. The visible light from the early Universe still remains, but in the form of microwaves.

The picture of the Universe shown in Figure 5.12 was established using data taken by the Planck satellite. This satellite scanned the Universe in search of this fossil radiation in the microwave range. This is the oldest picture that we have of the Universe, telling us how it looked while it was still in its infancy. It provides precious information about how particles managed to group together after the initial moments of the Universe. The first striking fact to note is that although the variations were quite small, the Universe was no longer uniform but was instead full of lumps. The different colors show that there were warmer spots corresponding to places where matter had already begun to agglutinate under the influence of the gravitational force.

Evolution of the Universe

We can analyze this cosmic radiation in the same way as we use a prism to decompose light into its various colors. Each color corresponds to a particular wavelength, possessing a very precise frequency. Cosmologists have studied the amount of radiation associated with each frequency. The various frequencies correspond to small variations in temperature, represented by small spots or lumps of different colors on the map in Figure 5.12. The size of each lump and its temperature are related to the evolution of the Universe.

The graph on Figure 5.13 was obtained from Figure 5.12 by plotting the temperature variation of each lump as a function of its size (or angular width) found from the photo. The points give the experimental results. The small vertical bar attached to each point represents the experimental error margin. These data are compared with predictions from a theoretical cosmological model (shown by the line) that describes how matter has formed and evolved in the Universe from the Big Bang until

Figure 5.12 Here is the oldest photo that we possess of the Universe, telling us today how it appeared 380,000 years after the Big Bang. Its content was no longer distributed uniformly but had already begun to agglutinate, forming lumps that served as "seeds" for galaxies. This photo has been reconstructed from the cosmic microwave background, which is radiation in the microwave range reaching us today from all directions in space. This radiation has been wandering around for approximately 13.8 billion years. As the Universe is essentially empty, nothing has hindered its propagation.

Source: Planck, European Space Agency.

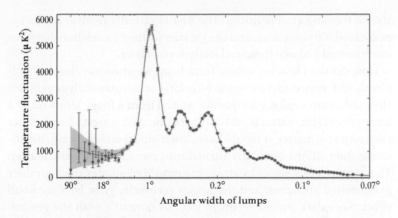

Figure 5.13 The temperature variation of each lump shown in different colors in Figure 5-12 is plotted here as a function of its size (or angular width). The solid line represents the theoretical cosmological model describing how the universe evolved from its beginning until now. Its six free parameters can be adjusted to fit the experimental data points. Two of these parameters are the density of dark matter and the density of dark energy in the Universe. Doing so, the Planck scientists determined that 27% of the content of the Universe is dark matter and 68%, dark energy.

Source: Planck, European Space Agency.

now. This model has six adjustable parameters, two of which are the density of dark matter and the density of dark energy. The Planck scientists determined these two densities by adjusting the parameters of the model to fit their experimental observations. This is how they determined that 27% of the content of the Universe is dark matter and 68%, dark energy.

Dark matter and seeds of galaxies

Cosmology, the science studying the evolution of the Universe, has confirmed the existence of dark matter not only by providing impressive agreement between the experimental data from Planck and theoretical predictions, but also by clarifying the essential role played by dark matter in the formation of galaxies. The vast majority of cosmologists now believe that all matter, both dark and visible, was nearly uniformly distributed just after the Big Bang, like a giant fog. As already mentioned, a fast expansion followed right after the Big Bang, allowing the Universe to cool down enough so that three minutes later, particles started forming atomic nuclei. The first electrically neutral atoms appeared 380,000 years later, and the galaxies formed somewhere between one hundred and one thousand million years later.

How did the Universe evolve from being an immense cloud of uniformly distributed matter to forming large structures such as galaxies? How did atoms coalesce so that we moved from a foggy Universe to a lumpy one? Dark matter is probably to blame. As it is most likely heavier than ordinary matter, it would have slowed down earlier. Small, microscopic fluctuations gradually turned into tiny lumps of dark matter. These lumps grew bigger by attracting more dark matter through their gravitational attraction and the lumps eventually grew by a snowball effect. Since dark matter seems to interact normally with the gravitational force but only very weakly or possibly not at all with the three other forces, these small accumulations of dark matter resisted better the storms of electromagnetic radiation present at the start of the Universe. In contrast, ordinary matter must have had a much harder time agglutinating in such a hostile environment.

Once visible matter cooled off after the expansion of the Universe, it too began to accumulate around the already formed lumps of dark matter. Consequently, dark matter sowed the seeds of galaxies. "All this would have been possible without dark matter, but it would have taken much more time," asserts Alexandre Arbey, a cosmologist working at CERN.

Simulating the formation of the Universe

To test such hypotheses, cosmologists use simulations. An evolution model must succeed in starting from the image that we have of the Universe 380,000 years after its birth, let it evolve for 13.8 billion years and see if it ends up giving something similar to what we observe today. Such models exist, and they can reproduce the evolution of the Universe in accelerated mode using computer simulations made possible by the tremendous computing power available today. There are several videos that illustrate this process: see, for example the short documentary "Formation of the Universe: The Big Computation," relating the work of Professor Jean-Michel Alimi's team at the National Center for Scientific Research (CNRS) in Paris (http://videotheque.cnrs.fr/), or this movie produced by the Planck experiment: http://www.wired.com/2014/05/supercomputers-simulate-the-universe-in-unprecedented-detail/. Both of these movies allow people in a hurry to relive the 13.8 billion years of evolution of the Universe and watch the formation of large-scale structures in a few seconds.

The images in Figure 5.14 give an idea of the results obtained with this technique, showing the structures contained in the Universe

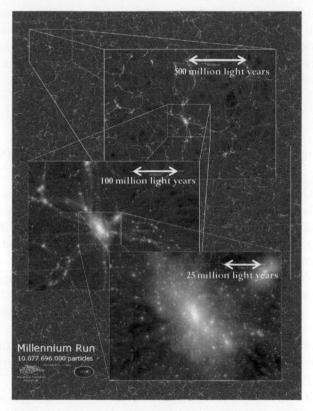

Figure 5.14 The distribution of matter in the Universe as obtained from computer simulations. The starting point was the oldest image of the Universe, obtained from the cosmic microwave background, showing the distribution of matter 380,000 years after the Big Bang. In the simulation, grains of matter were allowed to move under the influence of the gravitational force for 13.8 billion years (at accelerated speed, of course). The four photos show on various scales the structures the models predict should be found in the Universe today, the last three being zoomed in views. The predictions agree with current observations, which proves that the theoretical evolution models used, when the presence of dark matter is included, do indeed correspond to reality.

Source: Volker Springel and the Virgo Consortium.

reproduced by digital simulations. On the background picture, matter seems distributed almost uniformly but as soon as we zoom in on the image, large filamentous structures appear, as clearly seen in the second image. The brightest points correspond to places where dark matter is most concentrated and serve as seeds for the formation of galaxies. The most enlarged picture reveals galaxies as we are used to see them on images coming, for example, from the Hubble telescope. Theoretical models that do not include the presence of dark matter fail to reproduce these big structures, bringing in one more argument to support its existence.

Evidence supporting dark matter

To summarize, here is the evidence supporting the existence of dark matter:

1. The rotational speeds of stars in spiral galaxies indicate that these galaxies contain much more matter than what is visible.
2. Gravitational lenses reveal the presence of dark matter by diverting the light coming from celestial bodies located behind large lumps of dark matter.
3. Collisions of galactic clusters such as the Bullet Cluster captured by the Hubble Telescope clearly show that dark matter and ordinary matter act differently. The dark matter is revealed by gravitational lenses, and the visible matter by emitted X-rays.
4. Dark matter is an essential parameter needed to reproduce the distribution of lumps measured by the Planck experiment as shown in Figures 5.12 and 5.13.
5. Dark matter acted as a catalyst for the formation of galaxies, a phenomenon that would have taken much more time if only visible matter had been present.

Two hypotheses to discard

So dark matter exists, but what could it be? Nobody knows. I have often been asked if dark matter could be made of antimatter or black holes. Even though these two hypotheses might seem possible, that's really not the case. Here is why.

As we saw in the first chapter, matter and antimatter come in pairs and behave in more or less identical ways, even though no one knows why antimatter has essentially disappeared from the Universe today. Antimatter behaves like matter. For example, the positron is the antimatter counterpart of the electron. Just like the electron, the positron has an electrical charge and reacts to the electromagnetic force. The same thing goes for the antimuon, the antitau and the six antiquarks. Any electrical charge emits light when accelerated. Therefore, antimatter would shine and interact with ordinary matter. These characteristics eliminate it completely as a potential dark matter candidate.

Next hypothesis: could dark matter consist of black holes? To understand what a black hole is, one must first realize that an atom is essentially empty. Imagine an atom being magnified to fit a football field about 100 yards (or 100 m) long (Figure 5.15). The atomic nucleus, placed in the center of the field, would be the size of a die. Electrons would be sitting on the edge of the field. Most of the volume of an atom is thus just empty space.

Under certain conditions, massive stars sometimes collapse and start shrinking under the influence of their own gravitational attraction. Their atoms are then compressed to extremes, with their electrons

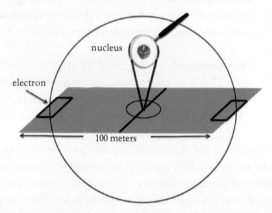

Figure 5.15 An atom is essentially empty space. If an atom was the size of a football field, the nucleus would be bigger than a die and the electrons would be on the edge of the field.

Source: Pauline Gagnon.

squeezed onto the nucleus. This is what leads to black holes. A star the size of our Sun—that is, 0.9 million miles (1.4 million kilometers) in diameter—would be reduced to a hyperdense object 2 miles (3 km) in diameter.

The gravitational field of such an object would be so strong that it would greatly distort the space around it, as we saw earlier, to the point where even light would not be able to escape from the black hole if it were to fall inside it, hence the name. There is one essential but much lesser known fact about black holes, though. Matter attracted by the strong gravitational field of a black hole will emit light during its acceleration toward the black hole. As long as this accelerated matter is a good distance away from the black hole, the emitted light will be diverted from its path but will still escape. Under these conditions, black holes do not have the characteristics of dark matter, since they emit a lot of radiation. That light would be detected.

Getting our hands on dark matter

Now that I have hopefully convinced you of the existence of dark matter, let us see how we might detect it directly. At present, all evidence of dark matter, although there is much such evidence and it is difficult to refute, is indirect. Dark matter is perceived solely through its gravitational and cosmological effects. Is there any more tangible and direct proof of its presence? This is what several teams of researchers are trying to establish, and there are heated debates over how to interpret the results.

Nobody has yet managed to observe dark matter directly and in an irrefutable way. This is not surprising, since we are speaking about matter of a completely different type that, unlike visible matter (we, all the planets, the stars and the galaxies), is not made of quarks or leptons.

Several hypotheses and theoretical models have been proposed to try to describe the nature of dark matter. One possibility consists of supposing the existence of particles that have no electrical charge and a very low mass but would interact with intense magnetic fields. These hypothetical particles are called *axions*. Two experiments at CERN, OSQAR and CAST, are currently trying to detect some of these to prove their existence by using powerful magnets. However, despite the great perseverance and

Table 5.1 Known and possible interactions of Standard Model particles and dark matter with the fundamental forces.

	Gravitational	Weak	Electromagnetic	Strong	Brout–Englert–Higgs field
Particles affected	All particles	Quarks, leptons	Charged particles	Quarks, gluons	Massive particles
Acts on dark matter?	Yes	???	No	No	???

ingenuity shown by the two teams of researchers, there are still no signs of the existence of axions.

Another, more widespread approach supposes that dark matter, just like visible matter, is also made of particles but these are different from axions or Standard Model particles. For us to be able to detect them, not only do they have to exist but they must also interact in one way or another with particles of ordinary matter.

As we saw in the first chapter, the fundamental particles of the Standard Model (the quarks and leptons) interact with each other through four different fundamental interactions, as well as with the Brout–Englert–Higgs field (Table 5.1). At present, all we know about dark matter is that it generates a gravitational field but does not interact through the electromagnetic force. Otherwise, it would emit light. If it interacted with the strong force, it would produce many interactions with ordinary matter. It would then be easy to detect, and numerous experiments would have already found it.

It would seem, then, that both the strong force and the electromagnetic force are excluded. But maybe dark matter interacts with ordinary matter through the weak force, the one responsible for radioactivity. If this hypothesis were correct, dark matter would then be made of weakly interacting particles. One other possibility, which I will explore at the end of this chapter, is the following. Since dark matter generates a gravitational field, it must have mass. If so, one might think that it would also interact with the Brout–Englert–Higgs field. These are some of the current hypotheses, and I will examine them in the following sections.

Looking for WIMPs

A very popular version of the first hypothesis in the previous paragraph suggests that dark matter particles might be weakly interacting massive particles, or WIMPs. The WIMPs would interact, although rarely, with matter just like neutrinos. A 20 lb (10 kg) detector would register less than one interaction per year with a dark matter particle. The exact number of collisions expected depends on the mass of the WIMPs, their abundance and their affinity for interacting with ordinary matter. None of these factors are known yet. To maximize our chances, people have built detectors containing as much material as possible—some using up to a ton of active material—to increase the odds of a collision between a WIMP and one atom of the detector.

Although this might seem incredibly small in comparison with the size of an LHC detector, which contains hundreds of tons of active material, these dark matter detectors are designed to provide an absolutely quiet environment. Imagine trying to detect the passage of a butterfly by looking for a minuscule ripple on the surface of a lake. The larger the surface of the lake, the more chance of detecting a passing butterfly. But this would only be achievable if the surface of the lake was perfectly still and sheltered from all types of perturbation. In comparison, an LHC detector is the equivalent of an agitated sea, open to winds, schools of fish and strong currents.

The Universe contains a huge quantity of dark matter. If dark matter can interact with regular matter, we can thus expect that from time to time, a WIMP will collide with a detector or, more exactly, with a proton or neutron in one of the atomic nuclei in the detector (Figure 5.16). Protons and neutrons are collectively called *nucleons*, since both particles are found in the nucleus. A collision between a WIMP and a nucleon would cause the atomic nucleus to recoil, inducing a small but detectable vibration (Figure 5.17).

The more voluminous a detector is and the longer we operate it, the more chances one has of registering a collision. And the more violent the shock that a WIMP causes to an atomic nucleus, the easier it will be to detect. Unfortunately, it is more likely that a WIMP will transfer just a small fraction of its energy to an atomic nucleus. So choosing the right material for a detector is not simple. For example, one would obtain more violent shocks, and thus more easily detectable collisions, with a detector made of germanium or silicon than with one built from heavier

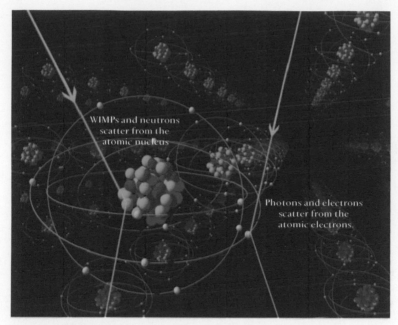

WIMPs and neutrons scatter from the atomic nucleus

Photons and electrons scatter from the atomic electrons

Figure 5.16 We assume that WIMPs, just like neutrons coming from cosmic rays, could collide with the protons and neutrons in the atomic nuclei of the detector material. Charged particles, such as electrons, interact with the electrons of the atoms, not the nuclei. These two types of interactions can be distinguished from each other.

Source: Mike Attisha for CDMS Collaboration.

nuclei such as xenon. On the other hand, the theory also predicts that the total number of collisions will be bigger for a xenon detector. The ideal detector does not exist: everything is a question of compromise, and this depends also on the efficiency of the possible techniques available for extracting these small signals from each material. Different teams have thus chosen to build detectors using different materials. In the end, this should pay off since one can check a broader array of possible scenarios, given that the exact characteristics of WIMPs are not known.

All of these detectors are installed deep in a mine or tunnel so that the overhead layer of rock acts as a screen to block incoming cosmic rays that would otherwise induce false signals in the detector. Eliminating all possible sources of background, such as cosmic rays and natural radioactivity, is the biggest challenge in these experiments.

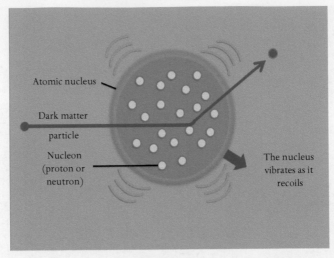

Figure 5.17 By striking a neutron or proton in the nucleus of one of the detector atoms, a WIMP would induce a small detectable vibration.

Source: Pauline Gagnon.

Dark matter rain

We know that there is dark matter at the centers of galaxies, since it acted as seeds for the galaxies, but it extends far beyond. The Earth should therefore be soaked in a mist of dark matter particles. Since the Earth is traveling around the Sun, this mist is similar to rain. The flow of dark particles would be of the order of six million particles per square inch (one million per square centimeter) per second if we suppose that WIMPs are ten times heavier than protons. This flow is enormous. Then, if these particles can interact with ordinary matter, even weakly, we ought to be able to captured some of them.

The principle is rather simple. Imagine a person standing on the deck of a cruise ship in a thick fog, without any wind. If the boat is at a standstill, the person will hardly get wet at all. But if the boat is moving through this thick fog made of fine droplets of water, the person will be sprayed with droplets. This effect will even be more pronounced if the person starts running. The person will receive more droplets of water when running in the same direction as the boat, and less in the opposite direction.

The same thing would occur for a particle detector operating on Earth. In June, the Earth's speed of rotation around the Sun, about 20 mi/s (30 km/s), lines up with the Sun's speed around the center of the Galaxy, 145 mi/s (235 km/s), increasing the spraying from the "WIMP rain" (Figure 5.18). In contrast, in December, the Earth's speed opposes that of the Sun, and the detector would meet fewer dark matter particles. A detector on Earth sensitive to WIMPs would therefore record more collisions in June than in December, since its speed relative to dark matter particles goes down from 165 mi/s (265 km/s) to 125 mi/s (200 km/s). These changes in the intensity of the WIMP rain would translate to a variation in the number of dark matter particles striking the detector over the year.

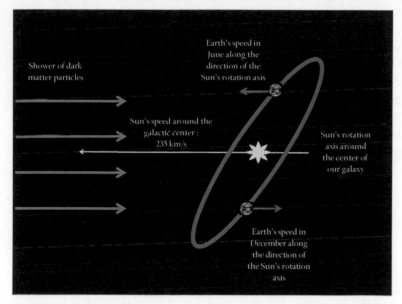

Figure 5.18 Because of the Earth's revolution around the Sun, the Earth's speed adds to the Sun's speed as the Sun moves around the center of the galaxy. The two speeds are aligned in June and opposed in December, as can be seen in the diagram. Imagine the WIMPs form a mist like droplets of water. The intensity of the "WIMP rain" striking the Earth will depend on the speed at which the Earth moves with respect to the rain. A detector on Earth will register fewer collisions with dark matter particles in December than in June, yielding an annual modulation of the signal.

Source: Pauline Gagnon.

This is exactly what scientists working on the DAMA/LIBRA experiment claim to have been observing for more than a decade now. Their signal is strong and clear: 8.9 sigma, that is, 8.9 times stronger than the possible statistical fluctuations, but unfortunately their claim is contradicted by several other experiments. The graph in Figure 5.19 shows the number of events registered by DAMA/LIBRA over time (more than 14 years in all). The annual modulation is clearly visible. It has become even more striking in recent years after the collaboration nearly tripled the size of their detector, making it even more sensitive. The DAMA/LIBRA team has not been the only one to make such a claim. Three other experiments have reported signals over the years: CoGeNT also detected a small annual modulation, while CRESST and CDMS observed more events than what could possibly have come from the background (cosmic rays, radioactivity etc.).

Everything would be great if these four experiments agreed. Unfortunately, this is not the case, as illustrated by the very messy plot in Figure 5.20. The complexity of the graph very adequately describes the current situation: we are swimming in total confusion. The vertical axis gives the effective cross section measured in square centimeters. This measures the target size, that is, how big nucleons (protons or neutrons) appear to a WIMP. This is how one measures the probability of an interaction between a dark matter particle and a particle of ordinary

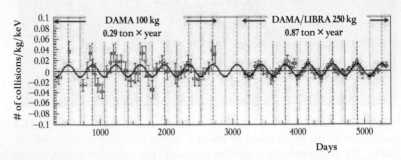

Figure 5.19 The annual modulation in the number of detected collisions registered by the DAMA/LIBRA detector. The team of researchers has attributed these results to collisions with dark matter particles but has not convinced the scientific community, since other experiments have not found any similar signals.

Source: DAMA/LIBRA.

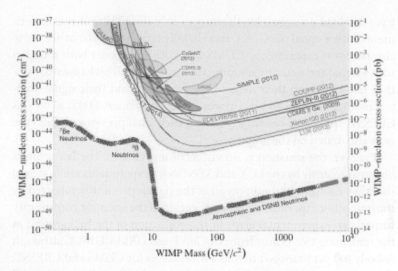

Figure 5.20 This very messy plot is here just to illustrate how confused the situation currently is in the search for dark matter. It summarizes the results of some (not even all!) of the current experiments that are looking for direct proof of the interaction of dark matter with ordinary matter. The vertical axis gives the probability of obtaining such interactions according to the hypothetical mass of the WIMPs shown on the horizontal axis. Four experiments (CoGeNT, DAMA/LIBRA, CDMS II and CRESST) have reported positive signals (shown by the closed colored zones), whereas several other groups (not all shown here) have excluded these four positive results as well as all values corresponding to the area above the open curves. See the text for more details.

Source: Julien Billard et al.

matter. The bigger the target seems, the easier it is to hit it. The horizontal axis simply gives the mass values in units of GeV for possible dark matter candidates.

This graph is in fact simpler than it appears at first glance. The closed areas shaded in various colors show the values obtained (and the associated error margins) by the four experiments that have reported a signal. The other, open curves give the exclusion limits measured by some of the numerous experiments that recorded no signal. Collectively, these combined results exclude the whole section in green, that is, roughly the top half of the graph. All values above these lines are excluded, which means that the results reported by the four groups that have claimed a signal are in complete contradiction with those of the experiments that have detected no signal. Furthermore, only the results of two experiments with a signal (CoGeNT and CDMS) agree with each other. Note also that two experiments, CDMS and CRESST, report both a region with a signal and a curve excluding the same signal. This happened after they had improved their detector performance and their signal analysis method. CDMS and its improved version SuperCDMS, as well as CRESST, have both recently excluded their own previous signals (the ones also drawn on the graph).

However, the situation is slowly becoming clearer. The latest limits obtained recently by the LUX and XENON100 experiments are so strong that they cast serious doubt on all of the signals previously reported by the four other experiments. A large portion of the scientific community suspects an experimental error in the evaluation of the background in the remaining two claims from CoGeNT and DAMA/LIBRA, although nobody has yet managed to find a mistake. As for CDMS and CRESST, their own most recent results supersede their previous observations. All of this illustrates how difficult these measurements are, as well as the progress being made to clarify the situation.

As frustrating as this situation may appear, it is far from surprising given the complexity of these experiments. Either we are dealing with experimental errors, or there is a theoretical explanation. Numerous theorists have made heroic efforts in the hope of constructing new models that could reconcile the fact that some experiments detect a signal whereas others don't. So far, no model has succeeded in achieving a consensus. Many experiments are continuing to accumulate data, and several others are under construction. Everybody is hard at work, on both the theoretical and the experimental fronts. We can expect a breakthrough in the coming years.

Unexplained signals from outer space

As we have just seen, there are several ongoing experiments actively trying to find irrefutable direct proof of the existence of dark matter particles. But the Earth is not the only place where physicists are looking. Several experiments conducted on board satellites (HEAT, Pamela and FERMI) and on the International Space Station (AMS-02) have reported, for several years now, an excess in the number of positrons (the antiparticle of electrons) seen in cosmic rays. The whole point is to understand where these positrons come from. As we saw in the first chapter, our Universe contains almost no antimatter. So what could be the origin of these positrons?

The graph in Figure 5.21 shows the fraction of positrons measured as a function of the energy for both electrons and positrons found in cosmic rays. This fraction was calculated with respect to all electrons and positrons found. The results of various experiments are shown, the most recent and most precise ones coming from the AMS-02 (dark circles), and PAMELA (open squares) experiments. The most intriguing fact is that the curve first goes up and then stabilizes around 200 GeV. The whole question now is to determine how this curve behaves at higher energy, to be able to sort out the question of the origin of these positrons.

Several possible explanations have been proposed. Here are the two most popular hypotheses. Some theorists suggest that these positrons could come from astronomical sources such as pulsars. These are neutron stars that spin on their axes and generate a pulsed signal owing to their powerful magnetic field. Others people think instead that this could be the first concrete sign of dark matter interacting with visible matter. Two dark matter particles could possibly annihilate with each

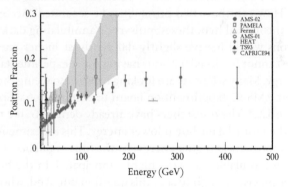

Figure 5.21 The fraction of positrons found in cosmic rays as a function of the energy of all electrons and positrons, as measured by several experiments. The latest results came from the AMS-02 experiment in September 2014 and are shown by the dots. One possible explanation is that these positrons could have come from the annihilation of dark matter particles, and this has drawn considerable attention from the scientific community. Many hope that AMS-02 will be able to sort this question out once they have accumulated and analyzed more data.

Source: AMS.

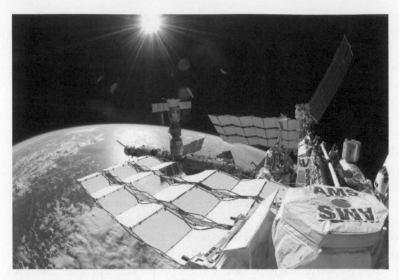

Figure 5.22 The AMS detector on board the International Space Station.
Source: AMS/NASA.

other and give off an electron and a positron, generating a source of
positrons. How can these two possibilities be separated? According to
these two theoretical hypotheses, pulsars or annihilating dark matter,
the positrons would behave slightly differently at higher energy. So,
this debate cannot be closed without having more experimental data at
higher energy. Many scientists are looking forward to seeing the results
of the latest AMS-02 experiment on board the International Space Station (Figure 5.22). The researchers have already demonstrated the high
quality of their initial data, but at lower energy. This experiment should
soon have more data at higher energy. At the time of publication, these
data were not available but were highly anticipated, in the hope that
they will be precise enough to sort this question out. And, who knows?
The data may possibly provide the first proof (although still indirect) of
an interaction between dark matter and ordinary matter. Lots of gray
matter is pondering over this.

Dark matter at the Large Hadron Collider

Underground and orbiting experiments have still provided no direct
proof of dark matter. One complementary but indirect way to look for it

is to use the ATLAS and CMS detectors at the Large Hadron Collider. We might indeed find dark matter particles there but, once again, if and only if dark matter interacts with some of the particles described by the Standard Model (namely, the fermions and bosons encountered in Chapter 1). As we do not know the exact process by which this might happen, physicists from the CMS and ATLAS experiments, working closely with theorists, must set all sorts of "traps," adapted to as many weird beasts as there are theories. With the restart of the LHC at higher energy, the hope is that one of the many approaches tested will reveal something new.

The most widespread idea is that there could be an extension to the Standard Model, a theory rooted in the principles of the Standard Model but going much further. One such hypothesis is called *supersymmetry* and the next chapter is dedicated to it. I will conclude this chapter by looking at some of the ways in which the LHC could reveal dark matter. We shall examine some other options suggested by supersymmetry in the next chapter.

How could dark matter be produced at the LHC?

Here is a description of some of the ways we think dark matter particles could be produced at the LHC. Each one comes in a multitude of variations. Hence, hundreds of physicists are working tirelessly on this question, doing their very best to leave no stone unturned. For example, the quarks and gluons contained in the colliding protons could produce both the known bosons of the Standard Model and new, hypothetical bosons. These new particles would come with various properties, none of which is known, so that we must test every possible value. And we must make further assumptions. For example, we may suppose that these bosons could in turn decay into a pair of dark matter particles, whose exact properties are equally unknown. All we know is that these particles must be electrically neutral. We do not know their mass, nor do we know which particles they could be produced from or by what mechanism. All these unknown quantities account for the multitude of possibilities that need to be checked.

Figure 5.23 shows a *Feynman diagram* illustrating one of these possibilities. One can see the incoming quarks (the lines labeled q) belonging to the colliding protons. Time flows from left to right. By convention, particles are represented by arrows pointing to the right, while antiparticles are represented by arrows pointing to the left, and so backwards in time. The energy released in the collision of the quarks materializes in the form of various bosons, represented generically by the symbols V and A, and ϕ/a. These can represent either known or hypothetical bosons. The lines associated with each class of particles are different to stress their differences. We use straight lines for fermions (quarks, leptons or dark matter particles, which we label χ), multilooped lines for gluons (denoted by g) and wavy lines for other bosons. The symbols g_q and g_{DM} represent the « couplings ». Roughly speaking, g_q indicates how likely two quarks could fuse to give a new boson V or A, and g_{DM} how likely two dark matter particles can emerge from such a boson.

The diagram in Figure 5.23 shows two colliding quarks combining to produce a boson of some sort, which then decays into a pair of dark matter particles, namely a particle and its antiparticle. In the case depicted here, a single gluon is also emitted by one of the incoming quarks. This is just like what could happen to a cyclist moving at too high a speed and losing his cap in a sharp turn. So out of the collision come one gluon and two dark matter particles. The gluon will pull out pairs of quarks and antiquarks out of the vacuum, which will form a bundle of hadrons (particles made of quarks). These bundles are called *jets*. The event will therefore contain one jet of particles and two dark matter particles.

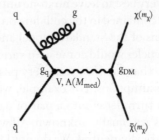

Figure 5.23 This "Feynman diagram" is a schematic representation of how particles are produced and how they decay. Time flows from left to right. Here we can see two quarks (q) coming in from the left, colliding and producing an intermediate state, a boson denoted V or A, which in turn is assumed to be able to decay into dark matter particles. This particular diagram illustrates one way dark matter particles could be produced at the LHC. These diagrams are all theoretical assumptions that need to be tested experimentally.

Source: Dark Matter Forum Report.

Seeing the invisible

As we saw in the second chapter, an event is a snapshot revealing how some heavy, unstable particles have decayed, producing several lighter, more stable particles. According to the principle of energy and momentum conservation, the energy and momentum must both be balanced in every event. If we observe the recoil of a rifle, a bullet must necessarily have left in the opposite direction. Likewise, a balloon, let free to go, produces thrust by forcing air out: the balloon pushes on the air and the air pushes the balloon in the opposite direction. The same thing holds for all the particles emerging from a collision: they must all be recoiling against each other. This is identical to what is seen with a firework: the fragments fly away in all directions, not in one single direction.

Figure 5.24 shows two events captured with the ATLAS detector. The image on the left shows a very common type of event, containing two jets of hadrons emerging from the decay of some heavier particle. The

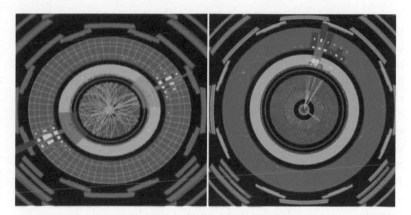

Figure 5.24 Left: a common type of event found in the ATLAS detector, containing two jets. Right: an extremely rare event, with a single jet, that could correspond to the signature of an invisible dark matter particle.
Source: ATLAS.

---

[1] Strictly speaking, energy is a *scalar* quantity, meaning it has no direction attached to it. On the other hand, a *vector* represents something that has a magnitude and a direction; for example, the velocity of a particle gives both its speed and its direction. However, in particle physics, we often ascribe the direction of the moving particle to the energy it carries.

two jets fly apart, recoiling against each other. The energy carried by all of the particles flying to the left is perfectly balanced by the energy carried by the particles flying to the right. The energy is balanced. There is no missing energy when all fragments are recorded.

Now look at the event on the right: a single jet is flying upward. But this jet must be recoiling against something that is flying downward, even though that "something" was not recorded in the detector and remains unseen. We can therefore conclude that something else was there. Hence, even when a particle leaves no signal in the detector, it can be "seen" thanks to the imbalance in energy in the event. This is how particles invisible to the detector, that is, particles that do not interact with the detector such as neutrinos and dark matter particles, can still be detected. And that is how we can see invisible particles.

Events containing single photons and missing energy

In the Feynman diagram shown earlier, a gluon was emitted by one of the incoming quarks. But photons can also be emitted that way. This is quite fortunate, since otherwise the only particles produced during the collision would be dark matter particles, and we would have no way to record such events in our detector. The photon makes this event detectable. This would look like the event in Figure 5.25, which was captured by the ATLAS experiment. We can spot the photon by its energy deposition, shown in yellow at about 4 o'clock in the left image and in the yellow tower in the right image, which represents the cylindrical part of the detector as if it had been unfolded. The pink dashed line around 10 o'clock represents the missing energy of an invisible particle recoiling against the photon. The blue lines correspond to particles produced in other low-energy collisions that occurred simultaneously, and can be ignored.

Unfortunately, other types of events collected by the ATLAS detector could look just like these events. They constitute the infamous background[2] that I have already mentioned in Chapter 4. For example, an event containing a Z boson and a photon would look just the same if the Z boson were to decay to two neutrinos (another type of particle

---

[2] The background corresponds to all types of events that have characteristics similar to the signal but come from other sources.

Figure 5.25 An event collected by the ATLAS detector containing a single photon (the yellow tower around the 4 o'clock position, also shown in the upper right window) and missing energy recoiling against the photon (the pink dashed line around 10 o'clock). This event has the characteristics of some invisible particles being produced in association with a photon, but since too few such events were found, it was attributed to background.

Source: ATLAS.

that does not interact with the detector, just like dark matter). In both cases, all we would see in the event would be a single photon and some missing energy. One relies on simulations and also on real data to evaluate the number of events coming from such background. For example, one can evaluate the number of events containing one photon and a Z boson by counting how many such events are found when the Z boson decays into two electrons or two muons. Since we know how often a Z boson decays into two neutrinos as opposed to two electrons or two muons, we can estimate this background. In the case of the event above, we concluded that it was compatible with coming from the background and not from a new type of invisible particle, since no events were found in excess of the expected number of background events.

Events with large missing energy

Here are some other ways one could produce dark matter particles at the LHC. In the left part of Figure 5.26, two colliding quarks produce one

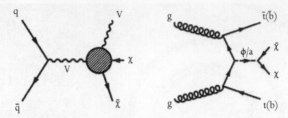

Figure 5.26 These "Feynman diagrams," just like the one shown earlier, give a schematic representation of how particles are produced and how they decay. Time flows from left to right. Here we see two particles come in, collide and produce other particles. These two particular diagrams illustrate some of the ways in which dark matter particles, denoted by the symbol χ could be produced at the LHC.

Source: Dark Matter Forum Report.

ordinary boson described by the Standard Model (denoted V): a photon, a Higgs, a Z or a W boson. The blob to the right simply represents our lack of knowledge about what might happen: some new, unknown type of interaction from which another boson emerges, as well as a pair of dark matter particles. Once again, the detector would record the signals left by these bosons or their remnants when they break apart, and a large amount of missing energy revealing the presence of unseen dark particles.

The diagram on the right of Figure 5.26 describes the case where two gluons produce pairs of *top* or *bottom* quarks, which in turn combine to give a new, hypothetical boson, ϕ/a. Once again, this boson could decay into a pair of dark matter particles. These two scenarios correspond to various hypotheses developed by theorists. Only experimental evidence will tell us if any one of them is correct.

All of these events containing dark matter particles, if they exist, would share one characteristic: they would all display an energy imbalance in the form of a large amount of missing energy. Such events would go unrecorded if it were not for the presence of one or sometimes two visible particles in the event. Searching for the presence of an unusual number of events containing a large amount of missing energy is essentially the strategy being adopted by CMS and ATLAS to look for dark matter particles at the LHC.

Could the Higgs boson be related to dark matter?

As mentioned earlier, since dark matter generates gravitational fields, it would seem that dark particles have a mass. If so, these particles may interact with the Brout–Englert–Higgs field. A Higgs boson should then be able to decay into dark matter particles. This is a possibility I spent many years exploring along with many other people.

At the LHC, Higgs bosons are sometimes produced in association with a Z boson. The left diagram in Figure 5.27 shows how this works. Two quarks q, belonging to two colliding protons in the LHC, can produce an excited Z boson (we indicate this excited state by using Z' in the diagram). This Z' returns to its normal state by shaking off the extra energy and emitting a Higgs boson (the H in the diagram). This is very much like what happens when an excited atom returns to its normal state by emitting a photon, and the reason why any material such as a piece of metal emits light when heated. When this occurs in the LHC, one ends up with a normal Z boson and a Higgs boson. Both can then decay into stable particles.

The Z boson will sometimes produce two leptons (two electrons or two muons). These are represented by the letter l in the diagram on the right in Figure 5.27, which shows which particles would pass through the detector. If we return to our initial hypothesis, our Higgs boson may

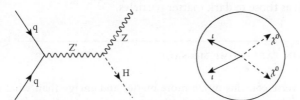

Figure 5.27 Diagram depicting how two quarks belonging to two colliding protons in the LHC can produce a Higgs boson accompanied by a Z boson. If the Higgs boson decays into dark matter particles, they will be invisible to the detector but will leave an imbalance in the energy of the event. Furthermore, the decay products of the Z boson (two muons or two electrons) will also be detectable. This is a particular signature that allows us to look for this type of event.

Source: Pauline Gagnon.

sometimes decay into two dark matter particles, denoted by the symbol χ° in the figure. In the end, only the fragments of the Z boson will be visible in the detector, not the decay products of the Higgs boson. Hence, the goal of such an analysis is to find events containing two leptons (electrons or muons) and some missing energy representing the two invisible particles.

ATLAS and CMS have examined closely all events collected in a search for events with these characteristics, but found nothing in excess of what was expected to come from the background. The main background in this case consists of events containing two Z bosons. The first one decays into two leptons and the other into two neutrinos, which are as invisible as dark matter particles. By using statistical methods similar to those described in Chapter 4 for the discovery of the Higgs boson, we had to conclude that nothing more than the background had been found. This allowed us, though, to set a limit on the probability of seeing dark matter particles interacting with ordinary matter.

This type of analysis at the LHC is sensitive even to very light dark matter particles. Remember the very messy diagram I showed summarizing some of the direct searches for dark matter in Figure 5.20? The CMS and ATLAS Collaborations can help to clarify the situation, even if their results depend on various theoretical hypotheses contrary to the results of direct searches. Efforts are ongoing and there is everything to hope for again since the restart of the Large Hadron Collider in spring 2015 at higher energy. More Higgs bosons are now being produced, increasing the chances of revealing even the most rare Higgs boson decays, such as those to dark matter particles.

THE MAIN TAKE-HOME MESSAGE

The Universe contains much more matter and energy than what is visible. The matter of all stars and galaxies accounts for only 5% of the total content of the Universe. The biggest share, 68%, appears to be in the form of some unknown type of energy and remains very mysterious. The rest, namely 27% of the Universe, consists of "dark matter," a type of matter that does not emit or absorb any light, hence its name. Dark matter seems to have very little in common with the fundamental particles of the Standard Model. However, its existence is in no doubt since we can detect its presence in many different ways through its gravitational effects. Dark matter is also an essential ingredient in the formation of galaxies. Without dark matter, cosmological models would be unable to reproduce the evolution of the Universe from the Big Bang 13.8 billion years ago to what is observed around us today.

Several experiments are in progress underground, in orbit around the Earth and at the Large Hadron Collider to detect dark matter particles. This will only be possible if dark matter interacts in some way with ordinary matter, and it is still not known if this is the case. A few experiments have reported having discovered dark matter particles, but several other experiments contradict those results. Much work is in progress, and we expect new developments soon (Figure 5.28).

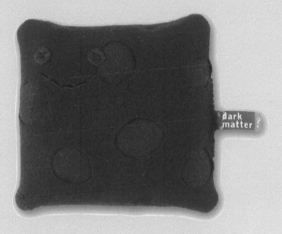

Figure 5.28 What will dark matter particles look like? Nobody knows. But here is how Julie Peasley, the Particle Zookeeper, imagines them.
Source: © Particle Zoo.

6

Calling SUSY to the Rescue

As we saw in the first chapter, the Standard Model describes the fundamental components of matter and the forces that ensure their cohesion. This model rests on two principles: first, all matter is made of particles, and second, these particles interact with each other by exchanging other particles associated with the fundamental forces. This model is both simple and very powerful since, of course, these two principles come with complex equations that describe in mathematical terms the interactions between particles. These equations allow theorists to predict, with extreme precision, which particles interact with others, how they decay and how often these decays occur. Until now, *nearly* every quantity that has been measured in particle physics laboratories over the last forty years has agreed perfectly with the value predicted by the theory, after taking the experimental error margins into account. Sometimes these predictions are exact up to the ninth decimal place. But this does not hold for all quantities.

That is the reason why theorists (such as John Ellis in Figure 6.1) know that the Standard Model is limited and that there has to be a more powerful and more encompassing theory that has not yet been discovered, in spite of the amazing success of the model. For example, as mentioned in the first chapter, the fact that neutrino masses are so small is already one clue telling us that the Standard Model does not explain everything.

The Standard Model is probably to particle physics what the four basic arithmetic operations (addition, multiplication, division and subtraction) are to mathematics. These four operations are enough to carry out the vast majority of our daily tasks, whereas for more complex calculations, one needs geometry, algebra and calculus. The Standard Model suffices to explain more or less all that has been observed until now, but it is probably only the visible part of the iceberg and the basis of a more sophisticated theory (Figure 6.2). Moreover, the

Figure 6.1 John Ellis, a theorist at CERN and King's College London and a strong supporter of supersymmetry, reading one of the many specialized articles published every day in physics journals, in an attempt to clarify the numerous unanswered questions.
Source: CERN.

model is now complete and predicts no new particles. Considerable efforts are thus being deployed to try to find new particles or faults in the model, since either could reveal a "secret passage" and put us on the right path to proceed further. Nevertheless, despite the numerous searches for new particles and despite having continuously improved the precision of our experimental and theoretical measurements, we have still not managed to find an anomaly or a new particle that would break the model and crack the door open toward what we call "new physics."

The Standard Model: a beautiful but flawed theory

What is wrong with the Standard Model? Why are there so many efforts at faulting it if all its predictions have turned out to be exactly right? Essentially, it leaves several points unanswered. For example, the model does not explain the asymmetry between matter and antimatter. Why does the Universe today consist essentially uniquely of matter, when matter and antimatter must have been produced in equal quantities

Figure 6.2 The Standard Model could be merely the tip of the iceberg. What more complete theory will explain the "new physics"?

Source: Anchor Inn, Twillingate, Newfoundland.

after the Big Bang? The Standard Model does not include gravity, one of the four fundamental forces. It does not explain either why the gravitational force is so much weaker than the three other forces, namely the electromagnetic, strong, and weak forces. For example, as we have already seen, a tiny magnet is powerful enough to thwart the gravitational attraction of the whole Earth and hold a small object on a refrigerator. Some theorists have proposed models with extra dimensions to explain the weakness of the gravitational force. As we saw in Chapter 5, a strong gravitational field deforms the space around it by bending it. But space can also contract. It could be that certain dimensions are completely bent, curled up on themselves, to the point of becoming microscopic. They would then be invisible to us. The strength of the gravitational force could be absorbed within one of these contracted dimensions, greatly weakening it. This would explain why gravity appears so weak, since we would only be observing the residue of a force originally as powerful as the others.

We live in a world with four dimensions: three dimensions of space and one of time. It might be possible that there are more dimensions

Figure 6.3 A tightrope walker can only move in one single dimension (forward or backward), whereas an ant can also walk around the cable, accessing an additional dimension.

Source: *Symmetry* Magazine.

but that they were hidden from us. Here is an example to illustrate this concept: think of a tightrope walker moving along a cable. From her viewpoint, there is only one possible dimension: she can only move forward or backward on the cable. She cannot move sideways, nor can she move up or down. On the other hand, an ant moving along the same cable can easily walk around it (Figure 6.3). For this ant, there are two dimensions, the second being bent around itself and nearly invisible on the human scale. These theories about extra dimensions predict the existence of new particles. If these hypotheses turn out to be true, these particles could be discovered at the LHC. Several searches are ongoing. In the meantime, nobody knows if these hypotheses are valid or not.

This enormous difference in the strength of the fundamental forces, characterized by the weakness of the gravitational force, is only one aspect of a more general problem called the "hierarchy problem." This term also refers to the wide range of mass values of the fundamental particles. To illustrate this, take a look at Figure 6.4. Some masses are expressed in units of electronvolts (eV), others in millions of eV (MeV) and others even in billions of eV (GeV). The electron (0.511 MeV) is 3500 times lighter than the tau (1.77 GeV). We can make the same observation for the quarks: the *top* quark, with a mass of 173.5 GeV, is 75,000 times heavier than the *up* quark (2.3 MeV). Why is there such a variety

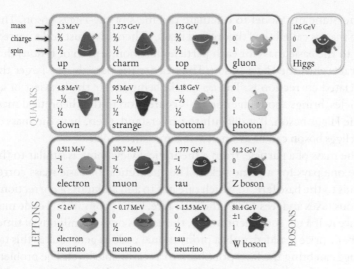

Figure 6.4 All of the known fundamental Standard Model particles together with their mass, their electrical charge and their spin. For the fermions, each column corresponds to a generation. Nobody knows why there is such a spread in mass values.

Source: Pauline Gagnon and © Particle Zoo.

of masses among the components of matter? Why do these particles come in three generations? All these questions are unanswered.

There is one other problem with the Standard Model: the grains of matter, the quarks and leptons, all have a spin of ½, and thus belong to the category of fermions as we saw in Chapter 1. The force carriers, on the other hand, have a whole-number value of spin as shown in Figure 6.4, and that means they are bosons. Why is this so? Why is there such a division between these two groups of particles? We do not know. As we saw in the first chapter, this distinction leads to completely different behaviors for these two types of particles, as if they belong to different worlds. Fermions have to obey strict exclusion rules, whereas bosons love company: the more the bosons, the merrier. We can pile an infinite number of identical bosons onto the same point in space, as in the case of superconductivity for example.

The hierarchy problem of the particle masses also affects the mass of the Higgs boson. The equations of the Standard Model relate fundamental particles to each other. For example, one can use the equations of the Standard Model to calculate the mass of the Higgs boson. This basic mass is called the "theoretical mass." Theorists must add a correction to the basic mass for each particle (fermion, lepton or boson) that interacts with the Higgs boson. The heavier the particle, the larger the associated correction is. The *top* quark, being by far the heaviest of all particles, brings about an enormous correction to the theoretical mass of the Higgs boson. It is difficult to understand, then, how the mass of the Higgs boson can be as small as it is measured to be.

The mass of a particle, from a theoretical viewpoint, is similar to the price one pays for a plane ticket. The theoretical or basic mass corresponds to the base fare, to which one has to add all kinds of corrections. Various taxes and fees add to the basic fare, but a promotional code may reduce it. If a tax equivalent to, say, thousands or even millions of times the basic price is added, only a promotional code as generous as this tax is huge can bring the final price back to a reasonable level. The problem with the calculated mass of the Higgs boson is that it undergoes wild fluctuations due to the correction coming from the *top* quark. This can only be cancelled out if there are new particles able to neutralize this contribution. For example, the correction to the Higgs boson mass from the *top* quark could be counterbalanced by equally large corrections coming from some as yet unknown particles. This would then explain why the mass of the Higgs boson is as small as what has been measured.

As a last resort, theorists could stretch the model and force several parameters to take very precise values to solve this problem of the theoretical mass of the Higgs boson. But this would be equivalent to trying to sew a dress with not quite enough fabric when the pattern called for much more. One would need to adjust all the pieces absolutely perfectly, saving every tiny scrap of fabric and piecing them together again to have enough fabric. This fine adjustment of the parameters of the theory by hand is called "fine tuning." Theorists loathe this kind of approach, considering it highly improbable and unnatural, and so they much prefer to avoid it at all costs.

If all these arguments have not yet convinced you, here is the one major reason of the inevitability of a more complete theory: the Standard Model describes only ordinary matter, the type one finds on Earth and in all stars and galaxies, as we saw in the previous chapter. Evidence abounds to indicate that the Universe contains five times more dark matter—a type of matter completely different from the type we know—than ordinary matter. Among all the fundamental particles described by the Standard Model, none possesses the properties of dark matter. It is thus clear that this model gives an incomplete image of the content of the Universe.

Why do we need new physics?

To summarize, the following are the main flaws of the Standard Model. Taken together, they suggest that there must be another, more encompassing theory describing something yet undiscovered and dubbed "new physics".

- There is no explanation for the smallness of neutrino masses and no answer to the question of whether or not the neutrino is its own antiparticle.
- The model does not explain the asymmetry between matter and antimatter (the near absence of antimatter in the Universe).
- It contains no particle that has the properties of dark matter.
- It does not include gravitation.
- It does not explain the weakness of the gravitational force.
- It does not explain the existence of three generations of particles, nor why their masses are so ill-assorted.
- It does not explain the division between fermions and bosons.
- It does not solve the problem of the theoretical mass of the Higgs boson.

All these reasons have led theorists to try developing a more complete theory for several years. This theory needs to be based on the Standard Model but solve at least some, if not all, of these problems. One of the proposed theories is called *supersymmetry*.

Supersymmetry: a tempting theory

Supersymmetry is a theory that first appeared in the early 1970s as a mathematical symmetry in "string theory." The latter theory was itself developed to unify the four fundamental forces. Supersymmetry would explain the division between bosons and fermions. Over time, many people have contributed new elements so that today, supersymmetry is a promising theory, but not the only one, that goes beyond the Standard Model.

Two Russian theorists, D.V. Volkov and V.P. Akulov, were among the pioneers. Then, in 1973, Julius Wess and Bruno Zumino presented the first supersymmetric model in four dimensions, paving the way toward future developments. The following year, Pierre Fayet generalized the Brout–Englert–Higgs mechanism to supersymmetry and introduced "superpartners" to the Standard Model particles for the first time. This crucial step led to establishing a symmetry between bosons and fermions, hence the name supersymmetry, or SUSY to its close friends.

There are several supersymmetric models based on the principles of supersymmetry. They build on the Standard Model and associate one or several partners with each fundamental particle. Fermions get bosons for superpartners, and vice versa. This unifies the fundamental components of matter with the force carriers. Everything becomes more harmonious and more symmetric.

Supersymmetric particles are labeled with a tilde (~) over their symbol, as shown in Figure 6.5, which is extracted from the excellent movie *Particle Fever*. The names of the superpartners associated with fermions are obtained by adding an *s* in front of the name of their partner, to stress their supersymmetric character. The *sbottom* is associated with the *bottom* quark, and the *stau* with the *tau* lepton. The model thus comes along with a whole slew of new bosons, called *squarks* and *sleptons*.

The supersymmetric particles associated with the bosons of the Standard Model are the *gluinos, photinos, Winos, Zinos* (also called *Binos*),

Figure 6.5 Supersymmetry builds on the Standard Model and adds a whole slew of new particles to the known ones.

Source: Mark Levinson, *Particle Fever*.

gravitinos, and *Higgsinos*. They are all fermions. By "combining" the fermionic superpartners of the electroweak force carriers (namely, the *photinos, Winos* and *Zinos*) with the *Higgsino*, we obtain particles with an electrical charge called *charginos* and neutral particles called *neutralinos*. Supersymmetry also has five different Higgs bosons, as we shall soon see. To the zoo of Standard Model particles, SUSY thus adds a whole menagerie of supersymmetric particles. This is one big downside of this theory. The number of fundamental particles more than doubles, wandering farther away from any dream of simplification. One step forward, but two steps back.

There are two major upsides, though. First of all, the two superpartners of the *top* quark, the *stops*, can counterbalance the huge correction generated by the *top* quark that affects the theoretical mass of the Higgs boson. Second, and not of minor importance, the lightest supersymmetric particle possesses exactly the characteristics expected for dark matter particles if we suppose that a property called "R-parity" is conserved. This property is not new, since Standard Model particles conserve R-parity too.

R-parity works slightly like the transmission of an evil card from one player to another in some card games, such as the game of Hearts, also known as Black Lady, where players must avoid getting the Queen of Spades in their tricks. If one cannot get rid of this card by giving it to another person, one gets stuck with it and loses several points. Similarly, the conservation of R-parity implies that supersymmetric particles can only decay to at least one other supersymmetric particle. Consequently, the lightest supersymmetric particle, or LSP, the last one of the decay chain, cannot decay into any other particle and is therefore stable. It exists forever, exactly like dark matter particles. This particle, the LSP, could then be the much sought-after dark matter particle. It must be electrically neutral since, as we saw, dark matter particles cannot have an electric charge; otherwise dark matter would emit light. In several supersymmetric models, the ideal dark matter candidate is the lightest *neutralino*.

To summarize, supersymmetry was first seen as a way to bring about more harmony by unifying the fermions and bosons of the Standard Model. It solves some aspects of the hierarchy problem, such as that related to the theoretical mass of the Higgs boson. What is really remarkable, though, is that this new theory, developed for completely different reasons, could also solve the huge problem of dark matter since it predicts the existence of new particles that have exactly the characteristics of dark matter particles. This explains its popularity, since it kills two birds with one stone. Unfortunately, SUSY, even if discovered, will not be the final answer, since it still fails to unify all forces: it leaves gravitation aside, just like the Standard Model.

Has anybody seen my supersymmetric particles?

If supersymmetry is as miraculous as it appears, how come none of the many new supersymmetric particles it postulates have been found by now? There are several possible reasons, the simplest being of course that this theory is wrong and that there are no supersymmetric particles. If that is indeed the case, another theoretical solution will be needed to fix the problems of the Standard Model. A new experimental discovery would help nudge theorists in the right direction. Theory and experimental research always progress together, one constantly inspiring the other. Anyway, theorists are convinced that they need to find a new theory that goes beyond the current model, although they are still at a loss about how to do that.

Even though we have not yet found supersymmetric particles, SUSY remains a completely plausible hypothesis. Its particles could have escaped detection for various reasons. Maybe we experimentalists have not looked in the right place, or in the correct way. Or maybe the supersymmetric particles are too heavy and were out of the current reach of our accelerators. We now have more chance of discovering SUSY particles after the restart of the LHC in 2015 at higher energy (13 TeV instead of the energy of 8 TeV in 2012), and delivering much more data (Figure 6.6). And if we still do not find them, new limits will be set on them, which will help us focus on the remaining possibilities.

Many free parameters and several models

It is not easy to work with SUSY (and I am not thinking of anyone in particular). What is its main flaw? It contains numerous undefined parameters and comes in a variety of models. One of them, called the MSSM (for Minimal Supersymmetric Standard Model), has 105 free parameters. These parameters represent quantities such as the masses of the supersymmetric particles and their couplings, that is, quantities

Figure 6.6 A physicist taking her shift in the CMS control room. In spite of the impressive quantity of data collected so far by the ATLAS and CMS experiments, there are still no signs yet of the presence of supersymmetric particles.
Source: CERN.

related to how often they are produced and to the probability that they will decay into other particles. We thus have 105 parameters, each one being free to take any value it wants.

A parameter is a bit like a dimension. Imagine that we are trying to locate a group of lost hikers somewhere in the Alps. It would be necessary to check every "point" on a map (for example, every 10 yards or 10 meters) covering the whole territory. Thus, even in a two-dimensional space such as the surface of the Alps (that is, a space with two free parameters), if we do not know the exact latitude and longitude, there are a horribly large number of potential places to be checked. In a space with 105 dimensions, it is necessary to specify the values of 105 parameters to locate an object. Try to imagine how one could find such an object if we do not know any of these 105 parameters. The number of points to be checked in this volume becomes astronomical.

On the one hand, supersymmetry does not specify which values these quantities can take. On the other hand, as soon as these parameters are known, or as soon as we fix their values, the relations between

all particles are precisely determined. As we have measured none of
these parameters yet, the only sensible approach is to make educated
guesses, that is, assign the values we think are the most probable. Theorists thus impose reasonable constraints, as in for example deciding
to limit the search for the lost hikers to dry land, thus eliminating all
lakes in addition of all of the places already searched. This is exactly the
approach adopted: we limit the scope of searches for supersymmetric
particles by eliminating improbable places. Physicists have had to make
suppositions to reduce the search zone. This is how various supersymmetry models appeared. Every model represents an attempt to confine
the search zone based on different assumptions.

One subset of the MSSM is called the CMSSM, for "Constrained
MSSM." This was developed to leave only a handful of free parameters,
in order to greatly simplify the model. This was achieved at the cost of
making difficult choices to fix the values of several parameters using
various hypotheses. This amounts to giving up the search for our hikers
in large areas, such as the whole of Switzerland, by assuming the hikers
do not like cheese. The CMSSM is losing support, since experimental
results tend to exclude it. Techniques have also greatly evolved, and this
has led to the development of new models that take real characteristics
of the missing hikers into account instead of being built on the suppositions of the rescue workers.

Theoretical advances

The recent determination of the mass of the Higgs boson imposes new
constraints on the existing models and has thus played a decisive role
in the development of a new class of *phenomenological* models, referred
to as "pMSSM." As their name suggests, these models were developed
using existing experimental constraints, that is, observed phenomena,
since any theoretical model must reproduce the experimental data. The
105 parameters of the MSSM can be reduced to 19 or 20, depending on
the model. These models have the advantage of building on more solid
ground.

Several teams of theorists and experimentalists have combined all
recent and past results to determine which zones are still allowed in
the reduced but still vast 19- or 20-parameter space of the pMSSM models. To achieve this, they first generated a list of all possible points in
this multidimensional space. This corresponds to millions of millions

of allowed values for the masses and couplings of all the hypothetical supersymmetric particles. At this stage, the 19 or 20 parameters still had no values assigned to them. In our example of the search for the lost hikers, this corresponds to making a list of all possible locations, every 10 yards or 10 meters, covering the entire area of the Alps.

The second step was to impose all known experimental constraints to see which points among all these possibilities were still allowed. This was done using measurements of the characteristics of the Z, W and Higgs bosons, high-precision results from heavy-quark decays, cosmology and, of course, all direct searches for supersymmetric particles at the LHC and elsewhere, and dark matter searches in underground experiments. In our rescue party example, we would at this stage cross off our list every location where someone has already looked.

This technique has the disadvantage of requiring a huge number of calculations to test every single point of this 19- or 20-dimensional space but, in the end, it is easy to see where supersymmetric particles could still hide. And it works. With this method, several groups have been able to show that the class of very constrained SUSY models such as the CMSSM already mentioned are now highly disfavored, being confined to very few allowed parameter values. The strongest constraints come from the absence of direct discovery by the ATLAS and CMS experiments of squarks with masses beyond 1 TeV, that is, eight times the mass of the Higgs boson found in 2012. The same technique applied to the more recent pMSSM models indicates, on the other hand, that there is still plenty of parameter space allowed where one type or another of supersymmetry model can exist, although this space is also more limited.

Thanks to this technique that mixes experimental results and theoretical knowledge, we can now reduce an almost infinite number of possibilities to a rather small number, allowing experimentalists to focus their searches better. Furthermore, this technique has already practically eliminated several models that did not properly describe reality.

The signature of a supersymmetric particle at the LHC

How can the Large Hadron Collider help? As we saw in Chapter 3, large detectors are arranged around the accelerator and act like huge cameras, recording how the unstable particles produced break apart.

The resulting snapshots allow physicists to determine the origin, direction and energy of each fragment, so that they can reconstruct and identify the initial particle. The search for supersymmetric particles is very similar to the quest for dark matter particles described at the end of Chapter 5, the reason being that the lightest supersymmetric particle could turn out to be the much sought-after dark matter particle. This unique particle would be invisible to the detector.

The diagram in Figure 6.7 shows a typical chain of decays for a supersymmetric particle. If we assume that R-parity is conserved, then supersymmetric particles will always be produced in pairs at the LHC. Each one will break apart according to a decay chain similar to the one depicted here. Black lines represent SUSY particles, whereas Standard Model particles appear in gray. This cascade of decays ends with the lightest supersymmetric particle, assumed here to be a neutralino. Such events would not only display a large amount of missing energy, but also contain several ordinary particles that would be produced along

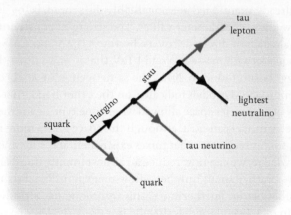

Figure 6.7 A typical decay chain for a supersymmetric particle. Every supersymmetric particle must decay into another, lighter supersymmetric particle (represented here by a black line) and an ordinary particle (shown in gray). The cascade of decays stops with the lightest supersymmetric particle (a neutralino in this example) if we assume that R-parity is conserved. Such events would therefore contain several ordinary particles plus a large amount of missing energy corresponding to escaping lightest supersymmetric particles.
Source: Fermilab.

the way and that would be visible to the LHC detectors. The main strategy is therefore to look for events that have lots of missing energy corresponding to invisible particles (as described in the previous chapter) and contain a few or several Standard Model particles, depending on the model. The number of variations on this theme is huge, each scenario corresponding to one particular hypothesis proposed by theorists. Hence, hundreds of physicists from ATLAS and CMS are exploring the many possibilities.

But suppose the colliding proton beams at the LHC only have enough energy to produce the lightest supersymmetric particles. Such events would contain nothing but invisible particles and there would be no way to detect them. We will only be able to catch them if these invisible particles are produced together with something else, such as if the incoming quarks or gluons emit a photon or a gluon as seen in Chapter 5 in the section "Dark matter at the Large Hadron Collider." One of the many strategies adopted by the CMS and ATLAS Collaborations in looking for supersymmetric particles consists of looking for events containing a lot of missing energy, corresponding to these invisible particles, plus other particles, such as a single jet of hadrons or a single photon.

Does this sound familiar? Indeed, such signatures would be very similar to those left by invisible dark matter particles, as we have seen in the previous chapter. If such events were found in excess of all background processes, they would reveal the existence of a new particle. The simplest way to determine whether this newly discovered invisible particle was supersymmetric or not would be to find other supersymmetric particles.

Is the Higgs boson a supersymmetric particle?

Now that we have measured its spin, we know that the new particle discovered in July 2012 is a Higgs boson. What kind of Higgs boson this is, exactly, is not yet known. Is this the unique boson predicted by the Standard Model, or are we dealing here with the lightest of the five Higgs bosons postulated by supersymmetry? This question will best be answered if we discover supersymmetric particles, since the lightest SUSY Higgs boson and the Standard Model Higgs boson have almost the same characteristics. In the meantime, all we can do is measure all the properties of this new boson with the highest possible precision. We have already established beyond any doubt that its spin is zero, a feature unique to Higgs bosons. We must still measure each of its decay channels with great accuracy and verify that every type of decay occurs exactly as predicted by the Standard Model. Any significant deviation could reveal a fault in the model.

Since its discovery, we have verified that the Higgs boson decays to bosons (photons, Z and W). In 2014, CMS and ATLAS provided evidence that the Higgs boson also decays to fermions, namely pairs made of a *b* quark and its antiquark, as well as pairs of *tau* and *antitau* leptons, based on all data collected up to the technical shutdown in 2013 (more on this in Chapter 10). The data accumulated at higher energy since 2015 is now used to refine these measurements and will help clarify the true nature of the Higgs boson. So far, all measurements made agree with the properties expected for the Standard Model Higgs boson. For example, we have measured the rates for the various decay channels and compared them with the values predicted by the Standard Model.

The graph in Figure 6.8 gives the values measured for several Higgs boson decay channels by the CMS experiment. Each point on the graph represents one channel, and five have been measured so far. A value of 1.0 means that we have measured exactly what is predicted by the theory. If the model is exact, we should then obtain a value of 1.0 for each decay channel. Taken individually or collectively, all of these channels in fact give a value compatible with 1.0 when we take the error margins into account. The combined value of 1.0 ± 0.13[1] corresponds to the average of these five channels, which happens by chance to fall right on the predicted theoretical value. We could equally well have obtained any value within the error margin of 0.13, represented by the gray strip. Similarly, the ATLAS Collaboration gets a combined value for all channels of 1.30 ± 0.19, also in agreement within the errors with the theoretical prediction of 1.0. However, the experimental error margins are still too large to make any definite statement, although there are no anomalies in sight. We need to accumulate and analyze way more events at higher energy

[1] The results quoted here were presented by CMS and ATLAS at the ICHEP conference held in Valencia, Spain, in July 2014.

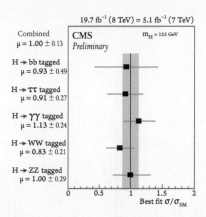

Figure 6.8 The results from the CMS Collaboration for the five Higgs boson decay channels as known in July 2014. Each channel would have a value of 1.0 if we were dealing with the Higgs boson predicted by the Standard Model. Indeed, all measurements agree with a value of 1.0, within the error margin shown by the gray strip.

Source: CMS.

before we can close this line of inquiry. This work is in progress, but will require all of the LHC data (and possibly a more powerful accelerator) to be fully completed.

Physicists from the CMS and ATLAS experiments have closely examined about 20 billion events collected by each experiment in the search for the possible appearance of supersymmetric particles, but to no avail. Both collaborations have tested tens of different approaches, and new possibilities are constantly being explored. Just to give you a taste of the kind of efforts made so far, take a look at the table in Figure 6.9. The first column lists about fifty different analyses that have been undertaken by the ATLAS Collaboration. CMS has achieved just as many. The horizontal green and blue bars show the excluded mass range for various supersymmetric particles, each one being targeted by a specific analysis. Even without looking at the details, one can quickly get the idea that if supersymmetry has not yet been discovered, it is not for lack of trying. But SUSY has not yet said its last word. The chances are still good that supersymmetric particles will appear now that the LHC has restarted in 2015. If this turns out to be the case, it would be as extraordinary as setting foot on a new inhabited planet.

ATLAS SUSY Searches* - 95% CL Lower Limits

ATLAS Preliminary

$\sqrt{s} = 7, 8$ TeV

	Model	e, μ, τ, γ	Jets	E_T^{miss}	$\int \mathcal{L} \, dt \, [\text{fb}^{-1}]$	Mass limit	Reference
Inclusive Searches	MSUGRA/CMSSM	$0\text{-}3\ e,\mu/1\text{-}2\ \tau$	2-10 jets/3 b	Yes	20.3	$\tilde{q}, \tilde{g}$ — 1.8 TeV	$m(\tilde{q})=m(\tilde{g})$	1507.05525
	$\tilde{q}\tilde{q}, \tilde{q}\rightarrow q\tilde{\chi}_1^0$	0	2-6 jets	Yes	20.3	$\tilde{q}$ — 850 GeV	$m(\tilde{\chi}_1^0)=0$ GeV, $m(1^{st} \text{ gen. } \tilde{q})=m(2^{nd} \text{ gen. } \tilde{q})$	1507.05525
	$\tilde{q}\tilde{q}, \tilde{q}\rightarrow q\tilde{\chi}_1^0$ (compressed)	mono-jet	1-3 jets	Yes	20.3	100-440 GeV	$m(\tilde{q})\text{-}m(\tilde{\chi}_1^0)<10$ GeV	1507.05525
	$\tilde{q}\tilde{q}, \tilde{q}\rightarrow q\tilde{\chi}_2^0 (\ell\ell/\ell\nu/\nu\nu)\tilde{\chi}_1^0$	$2\ e,\mu$ (off-Z)	2-6 jets	Yes	20.3	$\tilde{q}$ — 780 GeV	$m(\tilde{\chi}_1^0)=0$ GeV	1503.03290
	$\tilde{g}\tilde{g}, \tilde{g}\rightarrow q\bar{q}\tilde{\chi}_1^0$	0	2-6 jets	Yes	20.3	$\tilde{g}$ — 1.33 TeV	$m(\tilde{\chi}_1^0)=0$ GeV	1405.7875
	$\tilde{g}\tilde{g}, \tilde{g}\rightarrow qq\tilde{\chi}_1^\pm \rightarrow qqW^\pm\tilde{\chi}_1^0$	$0\text{-}1\ e,\mu$	2-6 jets	Yes	20	$\tilde{g}$ — 1.26 TeV	$m(\tilde{\chi}_1^0)<300$ GeV, $m(\tilde{\chi}_1^\pm)=0.5(m(\tilde{\chi}_1^0)+m(\tilde{g}))$	1507.05525
	$\tilde{g}\tilde{g}, \tilde{g}\rightarrow qq(\ell\ell/\ell\nu/\nu\nu)\tilde{\chi}_1^0$	$2\ e,\mu$	0-3 jets	—	20	$\tilde{g}$ — 1.32 TeV	$m(\tilde{\chi}_1^0)<300$ GeV	1501.03555
	GMSB ($\tilde{\ell}$ NLSP)	$1\text{-}2\ \tau + 0\text{-}1\ \ell$	0-2 jets	Yes	20.3	$\tilde{g}$ — 1.6 TeV	$\tan\beta>20$	1407.0603
	GGM (bino NLSP)	$2\ \gamma$	—	Yes	20.3	$\tilde{g}$ — 1.29 TeV	$c\tau$(NLSP)<0.1 mm	1507.05493
	GGM (higgsino-bino NLSP)	γ	$1\ b$	Yes	20.3	$\tilde{g}$ — 1.3 TeV	$m(\tilde{\chi}_1^0)<900$ GeV, $c\tau$(NLSP)<0.1 mm, $\mu<0$	1507.05493
	GGM (higgsino-bino NLSP)	γ	2 jets	Yes	20.3	$\tilde{g}$ — 1.25 TeV	$m(\tilde{\chi}_1^0)<850$ GeV, $c\tau$(NLSP)<0.1 mm, $\mu>0$	1507.05493
	GGM (higgsino NLSP)	$2\ e,\mu$ (Z)	2 jets	Yes	20.3	$\tilde{g}$ — 850 GeV	m(NLSP)>430 GeV	1503.03290
	Gravitino LSP	0	mono-jet	Yes	20.3	$F^{1/2}$ scale — 865 GeV	$m(\tilde{G})>1.8\times10^{-4}$ eV, $m(\tilde{g})=m(\tilde{q})=1.5$ TeV	1502.01518
3rd gen. $\tilde{g}$ med.	$\tilde{g}\tilde{g}, \tilde{g}\rightarrow b\bar{b}\tilde{\chi}_1^0$	0	$3\ b$	Yes	20.1	$\tilde{g}$ — 1.25 TeV	$m(\tilde{\chi}_1^0)<400$ GeV	1407.0600
	$\tilde{g}\tilde{g}, \tilde{g}\rightarrow t\bar{t}\tilde{\chi}_1^0$	0	7-10 jets	Yes	20.3	$\tilde{g}$ — 1.1 TeV	$m(\tilde{\chi}_1^0)<350$ GeV	1308.1841
	$\tilde{g}\tilde{g}, \tilde{g}\rightarrow t\bar{t}\tilde{\chi}_1^0$	$0\text{-}1\ e,\mu$	$3\ b$	Yes	20.1	$\tilde{g}$ — 1.34 TeV	$m(\tilde{\chi}_1^0)<400$ GeV	1407.0600
	$\tilde{g}\tilde{g}, \tilde{g}\rightarrow b\bar{t}\tilde{\chi}_1^\pm$	$0\text{-}1\ e,\mu$	$3\ b$	Yes	20.1	$\tilde{g}$ — 1.3 TeV	$m(\tilde{\chi}_1^0)<300$ GeV	1407.0600
3rd gen. squarks direct production	$\tilde{b}_1\tilde{b}_1, \tilde{b}_1\rightarrow b\tilde{\chi}_1^0$	0	$2\ b$	Yes	20.1	$\tilde{b}_1$ 100-620 GeV	$m(\tilde{\chi}_1^0)<90$ GeV	1308.2631
	$\tilde{b}_1\tilde{b}_1, \tilde{b}_1\rightarrow t\tilde{\chi}_1^\pm$	$2\ e,\mu$ (SS)	$0\text{-}3\ b$	Yes	20.3	$\tilde{b}_1$ 275-440 GeV	$m(\tilde{\chi}_1^\pm)=2 m(\tilde{\chi}_1^0)$	1404.2500
	$\tilde{t}_1\tilde{t}_1, \tilde{t}_1\rightarrow b\tilde{\chi}_1^\pm$	$1\text{-}2\ e,\mu$	$1\text{-}2\ b$	Yes	4.7/20.3	$\tilde{t}_1$ 110-167 GeV 230-460 GeV	$m(\tilde{\chi}_1^\pm)=2 m(\tilde{\chi}_1^0)$, $m(\tilde{\chi}_1^0)=55$ GeV	1209.2102, 1407.0583
	$\tilde{t}_1\tilde{t}_1, \tilde{t}_1\rightarrow Wb\tilde{\chi}_1^0$ or $t\tilde{\chi}_1^0$	$0\text{-}2\ e,\mu$	$0\text{-}2$ jets/1-2 b	Yes	20.3	$\tilde{t}_1$ 90-191 GeV 210-700 GeV	$m(\tilde{\chi}_1^0)=1$ GeV	1506.08616
	$\tilde{t}_1\tilde{t}_1, \tilde{t}_1\rightarrow c\tilde{\chi}_1^0$	0	mono-jet/c-tag	Yes	20.3	$\tilde{t}_1$ 90-240 GeV	$m(\tilde{t}_1)\text{-}m(\tilde{\chi}_1^0)<85$ GeV	1407.0608
	$\tilde{t}_1\tilde{t}_1$ (natural GMSB)	$2\ e,\mu$ (Z)	$1\ b$	Yes	20.3	$\tilde{t}_1$ 150-580 GeV	$m(\tilde{\chi}_1^0)>150$ GeV	1403.5222
	$\tilde{t}_2\tilde{t}_2, \tilde{t}_2\rightarrow \tilde{t}_1 + Z$	$3\ e,\mu$ (Z)	$1\ b$	Yes	20.3	$\tilde{t}_2$ 290-600 GeV	$m(\tilde{\chi}_1^0)<200$ GeV	1403.5222
EW direct	$\tilde{\ell}_{L,R}\tilde{\ell}_{L,R}, \tilde{\ell}\rightarrow \ell\tilde{\chi}_1^0$	$2\ e,\mu$	0	Yes	20.3	$\tilde{\ell}$ 90-325 GeV	$m(\tilde{\chi}_1^0)=0$ GeV	1403.5294
	$\tilde{\chi}_1^+\tilde{\chi}_1^-, \tilde{\chi}_1^+\rightarrow \tilde{\ell}\nu(\ell\tilde{\nu})$	$2\ e,\mu$	0	Yes	20.3	$\tilde{\chi}_1^\pm$ 140-465 GeV	$m(\tilde{\chi}_1^0)=0$ GeV, $m(\tilde{\ell},\tilde{\nu})=0.5(m(\tilde{\chi}_1^\pm)+m(\tilde{\chi}_1^0))$	1403.5294
	$\tilde{\chi}_1^+\tilde{\chi}_1^-, \tilde{\chi}_1^+\rightarrow \tilde{\tau}\nu(\tau\tilde{\nu})$	$2\ \tau$	—	Yes	20.3	100-350 GeV	$m(\tilde{\chi}_1^0)=0$ GeV, $m(\tilde{\tau},\tilde{\nu})=0.5(m(\tilde{\chi}_1^\pm)+m(\tilde{\chi}_1^0))$	1407.0350
	$\tilde{\chi}_1^\pm\tilde{\chi}_2^0\rightarrow \tilde{\ell}_L\nu\tilde{\ell}_L\ell(\tilde{\nu}\nu), \ell\tilde{\nu}\tilde{\ell}_L\ell(\tilde{\nu}\nu)$	$3\ e,\mu$	0	Yes	20.3	$\tilde{\chi}_2^0$ — 700 GeV	$m(\tilde{\chi}_1^\pm)=m(\tilde{\chi}_2^0)$, $m(\tilde{\chi}_1^0)=0$, $m(\tilde{\ell},\tilde{\nu})=0.5(m(\tilde{\chi}_1^\pm)+m(\tilde{\chi}_1^0))$	1402.7029
	$\tilde{\chi}_1^\pm\tilde{\chi}_2^0\rightarrow W\tilde{\chi}_1^0 Z\tilde{\chi}_1^0$	$2\text{-}3\ e,\mu$	0-2 jets	Yes	20.3	$\tilde{\chi}_2^0$ 420 GeV	$m(\tilde{\chi}_1^\pm)=m(\tilde{\chi}_2^0)$, $m(\tilde{\chi}_1^0)=0$, sleptons decoupled	1403.5294, 1402.7029
	$\tilde{\chi}_1^\pm\tilde{\chi}_2^0\rightarrow W\tilde{\chi}_1^0 h\tilde{\chi}_1^0, h\rightarrow b\bar{b}/WW/\tau\tau/\gamma\gamma$	e,μ,γ	0-2 b	Yes	20.3	$\tilde{\chi}_2^0$ 250 GeV	$m(\tilde{\chi}_1^\pm)=m(\tilde{\chi}_2^0)$, $m(\tilde{\chi}_1^0)=0$, sleptons decoupled	1501.07110
	$\tilde{\chi}_2^0\tilde{\chi}_3^0, \tilde{\chi}_{2,3}^0\rightarrow \tilde{\ell}_R\ell$	$4\ e,\mu$	0	Yes	20.3	$\tilde{\chi}_2^0, \tilde{\chi}_3^0$ 620 GeV	$m(\tilde{\chi}_2^0)=m(\tilde{\chi}_3^0)$, $m(\tilde{\chi}_1^0)=0$, $m(\tilde{\ell},\tilde{\nu})=0.5(m(\tilde{\chi}_2^0)+m(\tilde{\chi}_1^0))$	1405.5086
	GGM (wino NLSP) weak prod.	$1\ e,\mu + \gamma$	—	Yes	20.3	$\tilde{W}$ 124-361 GeV	$c\tau<1$ mm	1507.05493
Long-lived particles	Direct $\tilde{\chi}_1^+\tilde{\chi}_1^-$ prod., long-lived $\tilde{\chi}_1^\pm$	Disapp. trk	1 jet	Yes	20.3	$\tilde{\chi}_1^\pm$ 270 GeV	$m(\tilde{\chi}_1^\pm)\text{-}m(\tilde{\chi}_1^0)=160$ MeV, $\tau(\tilde{\chi}_1^\pm)=0.2$ ns	1310.3675
	Direct $\tilde{\chi}_1^+\tilde{\chi}_1^-$ prod., long-lived $\tilde{\chi}_1^\pm$	dE/dx trk	—	Yes	18.4	$\tilde{\chi}_1^\pm$ 482 GeV	$m(\tilde{\chi}_1^\pm)\text{-}m(\tilde{\chi}_1^0)=160$ MeV, $\tau(\tilde{\chi}_1^\pm)<15$ ns	1506.05332
	Stable, stopped $\tilde{g}$ R-hadron	0	1-5 jets	Yes	27.9	$\tilde{g}$ 832 GeV	$m(\tilde{\chi}_1^0)=100$ GeV, 10 $\mu s<c\tau(\tilde{g})<1000$ s	1310.6584
	Stable $\tilde{g}$ R-hadron	trk	—	—	19.1	$\tilde{g}$ 1.27 TeV		1411.6795
	GMSB, stable $\tilde{\tau}, \tilde{\chi}_1^0\rightarrow \tilde{\tau}(\tilde{e},\tilde{\mu})+\tau(e,\mu)$	1-2 μ	—	—	19.1	$\tilde{\tau}$ 537 GeV	10<$\tan\beta$<50	1411.6795
	GMSB, $\tilde{\chi}_1^0\rightarrow \gamma\tilde{G}$, long-lived $\tilde{\chi}_1^0$	$2\ \gamma$	—	Yes	20.3	$\tilde{\chi}_1^0$ 435 GeV	2<$\tau(\tilde{\chi}_1^0)<3$ ns, SPS8 model	1409.5542
	$\tilde{g}\tilde{g}, \tilde{\chi}_1^0\rightarrow eev/e\mu v/\mu\mu v$	displ. $ee/e\mu/\mu\mu$	—	—	20.3	$\tilde{\chi}_1^0$ 1.0 TeV	7 <$c\tau(\tilde{\chi}_1^0)<740$ mm, $m(\tilde{g})=1.3$ TeV	1504.05162
	GGM $\tilde{g}\tilde{g}, \tilde{\chi}_1^0\rightarrow Z\tilde{G}$	displ. vtx + jets	—	—	20.3	$\tilde{\chi}_1^0$ 1.0 TeV	6 <$c\tau(\tilde{\chi}_1^0)<480$ mm, $m(\tilde{g})=1.1$ TeV	1504.05162
RPV	LFV $pp\rightarrow \tilde{\nu}_\tau + X, \tilde{\nu}_\tau\rightarrow e\mu/e\tau/\mu\tau$	$e\mu, e\tau, \mu\tau$	—	—	20.3	$\tilde{\nu}_\tau$ — 1.7 TeV	$\lambda_{311}'=0.11, \lambda_{132/133/233}=0.07$	1503.04430
	Bilinear RPV CMSSM	$2\ e,\mu$ (SS)	$0\text{-}3\ b$	Yes	20.3	$\tilde{q}, \tilde{g}$ 1.35 TeV	$m(\tilde{q})=m(\tilde{g}), c\tau_{LSP}<1$ mm	1404.2500
	$\tilde{\chi}_1^+\tilde{\chi}_1^-, \tilde{\chi}_1^+\rightarrow W\tilde{\chi}_1^0, \tilde{\chi}_1^0\rightarrow ee\tilde{\nu}_\mu, e\mu\tilde{\nu}_e$	$4\ e,\mu$	—	Yes	20.3	$\tilde{\chi}_1^\pm$ 750 GeV	$m(\tilde{\chi}_1^0)>0.2\times m(\tilde{\chi}_1^\pm), \lambda_{121}\neq0$	1405.5086
	$\tilde{\chi}_1^+\tilde{\chi}_1^-, \tilde{\chi}_1^+\rightarrow W\tilde{\chi}_1^0, \tilde{\chi}_1^0\rightarrow \tau\tau\tilde{\nu}_e, e\tau\tilde{\nu}_\tau$	$3\ e,\mu + \tau$	—	Yes	20.3	$\tilde{\chi}_1^\pm$ 450 GeV	$m(\tilde{\chi}_1^0)>0.2\times m(\tilde{\chi}_1^\pm), \lambda_{133}\neq0$	1405.5086
	$\tilde{g}\rightarrow qqq$	0	6-7 jets	—	20.3	$\tilde{g}$ 917 GeV	BRI$(\tilde{g}\rightarrow$BRI$(b)\rightarrow$BRI$(c)>0\%$	1502.05686
	$\tilde{g}\tilde{g}, \tilde{g}\rightarrow q\tilde{\chi}_1^0, \tilde{\chi}_1^0\rightarrow qqq$	$2\ e,\mu$ (SS)	0-3 b	Yes	20.3	$\tilde{g}$ 870 GeV 650 GeV	$m(\tilde{\chi}_1^0)=800$ GeV	1404.250
	$\tilde{t}_1\tilde{t}_1, \tilde{t}_1\rightarrow bs$		2 jets + 2 b	—	20.3	100-308 GeV		ATLAS-CONF-2015-026
	$\tilde{t}_1\tilde{t}_1, \tilde{t}_1\rightarrow b\ell$	$2\ e,\mu$	$2\ b$	—	20.3	$\tilde{t}_1$ 0.4-1.0 TeV	BRI$(\tilde{t}_1\rightarrow be/b\mu)>20\%$	ATLAS-CONF-2015-015
Other	Scalar charm, $\tilde{c}\rightarrow c\tilde{\chi}_1^0$	0	$2\ c$	Yes	20.3	$\tilde{c}$ 490 GeV	$m(\tilde{\chi}_1^0)<200$ GeV	1501.01325

$\sqrt{s} = 7$ TeV	$\sqrt{s} = 8$ TeV

Mass scale [TeV] 10^{-1} 1

*Only a selection of the available mass limits on new states or phenomena is shown. All limits quoted are observed minus 1σ theoretical signal cross section uncertainty.

Figure 6.9 A long but not exhaustive list of analyses undertaken in search of supersymmetric particles. None of the 50 scenarios studied by ATLAS revealed the presence of new particles. The mass values excluded so far for these particles are given.

Source: ATLAS.

Parallel worlds

A group of very serious theorists have developed a surprising theory about dark matter (the original article can be found at http://arxiv.org/abs/0810.0713). This incorporates the idea of a "Hidden Valley" separating two worlds evolving in parallel. (I am not making this up: see http://arxiv.org/abs/hep-ph/0604261.pdf). On the one hand, there would be the material world that we know, containing all the Standard Model particles as well as those postulated by supersymmetry (although these are still hypothetical). On the other hand, there would also be a parallel "dark world" populated with dark matter particles as shown in Figure 6.10. The vertical axis gives the masses of the particles and each horizontal line represents a particle of a given mass. The heaviest ones are thus placed above the lightest ones.

It is possible to test this idea at the LHC, if we assume it can produce heavy supersymmetric particles. These particles would decay in cascade until they

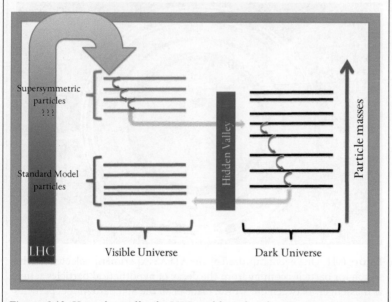

Figure 6.10 Hypothetically, the LHC could produce heavy supersymmetric particles. These would decay in cascade until the lightest one is produced. Some theories suggest that this particle could be some sort of a "messenger" capable of traveling between our world and another parallel dark Universe.
Source: Pauline Gagnon.

continued

Parallel worlds (*continued*)

reached the lightest supersymmetric particle. This one would be a "messenger" (the upper light gray horizontal arrow) capable of crossing the Hidden Valley as if going through a tunnel. It would then escape into a parallel Universe, the dark sector, becoming invisible to us.

In the dark sector, this particle would no longer be the lightest allowed particle and could decay again in a cascade of dark particles until it produced the lightest of all of the dark supersymmetric particles. This particle would be another messenger (the lower light gray arrow), capable of crossing the Hidden Valley again. Back in our world, it could decay to produce numerous pairs of light particles such as electrons and muons. This theory clearly establishes beyond any doubt that the particle physics world has no need to envy anything from science fiction.

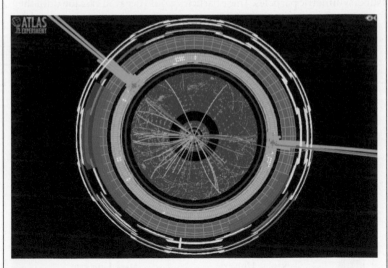

Figure 6.11 One event captured by the ATLAS experiment, selected for the search for particles coming from the decay of hypothetical particles acting as messengers between our world, made of all the Standard Model particles and supersymmetric ones, and another parallel world, made of dark matter. This messenger could decay so as to give off jets of energetic electrons or muons. No excess of events was found, and this event most likely comes from some background process.

Source: ATLAS.

Parallel worlds (*continued*)

Until recently, I was one of the experimentalists looking for signs of this Hidden Valley at ATLAS. We selected events containing pairs of regrouped electrons and muons. Unfortunately, we found nothing exceeding the expected background level. The image in Figure 6.11 of an event collected by the ATLAS detector shows the kind of signature that we were looking for: electrons (shown by green lines) produced in groups and carrying a lot of energy. Their tracks would be nearly straight, since at high energy they move too fast and the magnets cannot bend their trajectories.

Searches are ongoing there and in many other (maybe more likely) places. We are constantly refining our analysis methods and developing new strategies. If dark matter interacts with ordinary matter, we will eventually find it.

THE MAIN TAKE-HOME MESSAGE

The Standard Model, in spite of its highly precise predictions and its undeniable success, is riddled with holes. The most notable ones are the absence of particles corresponding to dark matter and the asymmetry between matter and antimatter in the Universe, that is, the unexplained disappearance of antimatter. Theorists thus know that there has to be a more encompassing theory. One popular theory that would allow us go further and reach what we call the "new physics" is supersymmetry, or SUSY. This solves some problems of the Standard Model and, as a bonus, predicts the existence of new particles, one of which has the characteristics of dark matter. But all is not perfect with SUSY either, its biggest flaw being that it has still not been discovered. None of the new SUSY particles have been found despite the considerable effort that has been made, although there are still numerous unexplored possibilities. Of course, if these particles are very heavy, they would have been out of the reach of the Large Hadron Collider when it operated at an energy of 8 TeV. We can only hope for exciting surprises now that the LHC has restarted in 2015 at higher energy, namely 13 TeV.

7

What Does Basic Research Put on Our Plates?

Basic research in particle physics is certainly fascinating (or so I hope to have shown you), but it comes at a price. For example, the construction cost of the Large Hadron Collider at CERN (personnel, machine R&D and materials) was about 3 billion euros (that is, roughly 3.3 billion US dollars). The construction of the ATLAS detector itself cost 455 million euros (500 million US dollars). As high as these figures may appear, though, CERN's annual budget of 825 million euros (approximately 900 million US dollars) is only equivalent to the cost of a cup of coffee for every European citizen old enough to drink one.

These are still huge amounts of money, so everyone is allowed to ask if it is worth it. In this chapter, I will show how money invested in research not only generates economic returns a hundredfold greater but also benefits society as a whole. Thanks to the technological breakthroughs that come from research, medical techniques and communication technology have improved. Basic research in physics has completely changed the way we live, and continues to do so.

Throughout this chapter, I will mostly give examples from CERN since it is the biggest international research laboratory for particle physics currently in operation. J-PARC in Japan, a multipurpose research center, also uses a proton accelerator. Other laboratories, such as SLAC and Fermilab in the United States and DESY in Germany, were until quite recently very active research centers for particle physics but their accelerators have now stopped operating. At Fermilab, however, the Main Injector is still running, providing neutrino beams for MINOS, Minerva and NOVA. Other experiments are in the approval or construction phase. There are also several other smaller research centers, such as SNOLAB in Sudbury in Canada, KEK in Japan and Gran Sasso in Italy, all specializing in neutrino physics and dark matter searches. Recently, all of the countries involved in particle physics research have decided to share their resources in large international collaborations such as those that are active at CERN.

The payoffs from fundamental research in particle physics are not necessarily direct. For example, nobody knows at the moment if the Higgs boson will ever have any practical use. Most likely it will not! Research is not conducted in the hope that the Higgs boson will solve the big problems of humanity. Instead, it is undertaken with the aim of understanding better the material world that surrounds us and pushing the state of knowledge one notch higher.

The first mission of a fundamental research laboratory is thus to fulfill the deep need human beings have for knowledge. Since the dawn of humanity, people have always wondered about their origins and their fate. But these laboratories also have three other main objectives: to contribute to technological development, to train a highly specialized workforce and, in the case of international laboratories, to promote peace and international collaboration through scientific research (Figure 7.1).

Figure 7.1 Every year, 250 students, from about 50 different countries, take part in the CERN Summer Student Program. Not only do they contribute to various research activities but they also meet young people from all over the world.

Source: CERN.

Nevertheless, one should not underestimate the potential of any new discovery. Who could have predicted a hundred years ago the incredible impact that research conducted by physicists on electrons and electromagnetic waves would have on our lives? One (disputed) anecdote illustrates this. When the British Chancellor of the Exchequer (minister of finance) questioned Michael Faraday about the potential utility of his research on electricity, he apparently answered that he did not know what could be made of it, but added "One day, sir, you may tax it."

Research on the electron and electromagnetism has led to the development of electronics, telecommunications and computers. Work undertaken by physicists in previous centuries, combined with the know-how of technicians and engineers who turned discoveries into practical applications, has reshaped our daily life. Without fundamental research in physics, we would still be reading by candlelight. We would certainly have very beautiful candles, but they would still only be candles, as a colleague pointed out to me. Not only does basic research have a major impact on our lives, but it also enlightens our spirits and frees humanity from the burden of ignorance.

Basic research, on both the theoretical and the experimental fronts, is guided by curiosity. It must be accomplished without constraints to let imagination and creativity flow freely. One must look everywhere, even when there is no assurance that something may be discovered. On the other hand, applied research is aimed at finding practical solutions to concrete problems. It takes advantage of technological breakthroughs ensuing from fundamental research activities and develops them further. The physical sciences are used by several other disciplines and play an essential role in various industrial sectors. From an economic point of view, physics influences the whole of society, and the returns in a vast array of domains impact every one of us in our daily life, as we shall see throughout this chapter.

Economic returns

Several studies have attempted to evaluate the impact that basic research has on the economy. One of them,[1] undertaken by Cebr (the Centre for Economics and Business Research) for the European Physical Society, is edifying. This estimates the impact of basic research on the whole sector

---

[1] This report is available online at http://www.eps.org/?page=policy_economy.

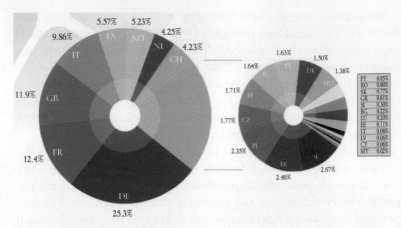

Figure 7.2 Percentage of total income for various European countries coming from the contribution of the industrial sector that depends on physics. Countries are identified by their two-letter codes: DE, Germany; FR, France; GB, Great Britain; IT, Italy; ES, Spain; NO, Norway; NL, Netherlands; CH, Switzerland; etc.

Source: European Physics Society.

of industries in Europe that depend on physics, from a technological and scientific viewpoint. It hence covers all economic activities depending on electrical, mechanical and civil engineering; energy; computing; communications; design and manufacturing; transport; medicine; and aeronautics.

In 2010, this sector generated 3.8 thousand billion euros in income for the 27 countries of the European Union, plus Switzerland and Norway (Figure 7.2). This corresponds to approximately 15% of the total income of these countries, and exceeds all retail sales. In total, 15.4 million people worked in this sector, that is, 13% of the total workforce of Europe.

Contributing to technological development

As we have seen throughout the book, research in particle physics today requires highly sophisticated tools. The technologies needed generally do not exist during the design stage of large experiments. They must be developed along the way, especially for huge projects such as the Large Hadron Collider, which was planned two decades in advance. The construction of the LHC allowed several technologies to be pushed beyond their existing boundaries. Never before had an instrument been built using such powerful superconducting magnets, not to mention the size of the project. Technologies related to superconductivity, extreme vacuum and extremely low temperatures have all greatly evolved.

The same goes for all of the measurement devices used by the large collaborations at the LHC for their detectors. All of these devices required more radiation-resistant, higher-performance electronic modules to withstand extreme radiation levels and cope with data collection at high speed and in huge volumes. This need provided the impetus to build the Grid, a vast computer network, connecting hundreds of thousands of computers scattered all over the world. The Grid supplies the computing power required by the LHC experiments.

This progress on the technological front has materialized into a myriad of industrial applications in domains as diverse as they are specialized. For example, these applications include humidity sensors equipped with optical fibers, a diaphragm system for engines that uses permanent magnets, open-source software for the design of printed circuit boards, and the additive processing technique used for 3D printing, to name but a few.

Some discoveries have also had a direct impact on the daily life of the vast majority of the planet's inhabitants. That is the case for the most phenomenal return from CERN, the World Wide Web. This has profoundly modified our access to information and knowledge, including in emerging countries, thereby affecting the day-to-day life of billions of individuals on Earth (see the box "The best gift from CERN").

The best gift from CERN

So far, the biggest impact of CERN on humanity has not been the discovery of the Higgs boson but rather the invention of the World Wide Web (WWW). Developed in 1989 by Tim Berners-Lee (Figure 7.3) and his team while he was working at CERN, the WWW was aimed at solving a problem that affected thousands of researchers at CERN. The scientists needed an effective way to exchange information. The majority of these physicists regularly commute between their institute and the laboratory to take part in research activities.

continued

The best gift from CERN (*continued*)

Figure 7.3 Tim Berners-Lee invented the World Wide Web while he was working at CERN. He is shown here in 1994 sitting in front of a computer screen displaying the very first Web page. According to some sources, the Web stimulates €1.5 trillion in annual commercial traffic.

Source: CERN.

The Web was thus born from their need to exchange information without having to lug kilos of printed documents in their suitcases.

But if Tim Berners-Lee was a visionary, CERN was also very forward-thinking in deciding to offer it to humanity without claiming any copyright. Since its research is subsidized by public money, CERN wanted to ensure that the Web would benefit everyone. Who could ignore the impact this communication tool has had on our lives, since it allows information to be distributed and accessed anywhere on the planet?

The particle physics community is adopting more and more an "open-source" approach, where knowledge is shared freely, at no cost, and distributed via the Web, for example. CERN experiments no longer publish their results exclusively in expensive, specialized journals but today, all information is available in open-source media. This applies not only to scientific publications but also to some software that is shared with other institutes, industry or society in a spirit of collaboration and sharing. This ensures that universities and institutes from emerging countries are not penalized.

Training a highly specialized workforce

All physics laboratories in the world participate in forming a highly specialized workforce. For example, besides its 2500 employees, CERN welcomes more than 13,000 researchers, PhD students, engineers and technicians to pursue research. They come from institutes attached to 78 countries and are of 112 different nationalities.

Every year, 1000 high school teachers from all continents come to discover CERN's facilities and research program. They can then transmit their acquired knowledge to their pupils. In 2014, CERN also started a new competition for high school students open to students from anywhere in the world. So far, teams from Greece, the Netherlands, Italy and South Africa were invited to come to the laboratory to conduct their own experiment. Students, researchers and technical staff come regularly for different periods of time for internships. Each summer, approximately 250 students from around the world participate in CERN's summer program. These young people receive a broad education in addition to contributing to the actual research program. Various physics laboratories also sponsor summer schools for advanced students on accelerators, physics and computing, as part of their training mission.

Only a fraction of the students trained in particle physics pursue a career in research, whether by choice or not, as the number of positions available is very limited. These highly qualified people move into a variety of fields, including finance, industry, communications, and computing. Some physicists at the early retirement stage even go as far as writing popular science books!

All physics laboratories happily open their doors freely to the public. CERN alone welcomes over 100,000 visitors every year. For example, in 2013, people from 63 different countries came for a visit; 40% of them were students. One might conclude that CERN really likes visitors, since it also organized "open days" on September 28 and 29, 2013 (Figures 7.4 and 7.5), attracting an additional 70,000 people from all continents who made the trip for the occasion. Fortunately, 2300 people volunteered to welcome them, taking 20,000 of them underground to explore the facilities. The others could choose from among 40 activities, either to have discussions with researchers, visit various workshops, attend short conferences or to discover a universe as fascinating as it is varied. These open days are organized roughly every 5 years, so keep your eyes open

Figure 7.4 Visitors discovering the CMS detector during CERN Open Day in September 2013.

Source: CERN.

Figure 7.5 During the open days, 70,000 people descended upon CERN as here, in a cryogenics laboratory where temperatures near absolute zero are reached. Visitors could try this small superconducting "scooter" that moves without friction by levitating over rails.

Source: CERN.

for the next ones. This is an opportunity not to be missed. It is also possible to visit CERN at any time by contacting the Visits Service.[2]

Promoting peace and international collaboration

Today, practically all experiments in particle physics are conducted by international teams. The scale of these research projects requires such collaboration and the pooling of resources. CERN is famous for this, since this spirit of collaboration appeared there long before the projects became gigantic. It was created in 1954 under the aegis of UNESCO, "in the aftermath of the Second World War, when Europe was in ruin, when absolutely everything had to be rebuilt." Some scientists and diplomats, including François de Rose (Figure 7.6), "understood the importance of

Figure 7.6 On the eve of his 103rd birthday, François de Rose, a former French ambassador and one of the founders of CERN, visited the ATLAS experiment and said that he was proud of the lab, which he described as a "huge European success." He died in March 2014 just after he had published his memoir, "A diplomat in the century."
Source: CERN.

2 http://outreach.web.cern.ch/outreach/visites/index.html

reviving fundamental research and, above all, of cooperation on a continental scale as the driving force of this ambition," as my colleague Corinne Pralavorio wrote in a eulogy for François de Rose.[3]

Twelve founding states[4] answered this rallying call. Sixty years later, CERN now has 23 member states. Overall, 78 different countries send scientists to CERN to take part in its research. In recent years, CERN has made a major change in direction. It is now recruiting new members not only from Europe as used to be the case, but also from all continents. Israel and Serbia were the first countries to join after this opening up, while Pakistan and Turkey are now Associate Members. Cyprus and Slovenia have also expressed their intention to become official members. Meanwhile, India, Japan, Russia and the United States already have observer status and could become full members. Member states contribute part of their gross national product to CERN's budget and decide on its functioning through their representatives at the CERN Council, the governing body of the laboratory. The nonmember states contribute financially to various research projects in which their scientists are involved.

The experiments in particle physics are thus models of international collaboration. It is not unusual to find scientists who come from countries that have no diplomatic relations with each other but who are working on the same project toward a common goal. CERN works so well that it served as a model for starting a similar laboratory in the Middle East. The SESAME project (Synchrotron Light for Experimental Science and Applications in the Middle East), is a multidisciplinary research center currently under construction in Jordan. It will recruit researchers from the region, including Palestine, Israel and Pakistan. Some of its scientists were trained at CERN (Figure 7.7).

Knowledge transfer

Basic research drives innovation. Particle physics experiments require state-of-the-art technology and are constantly pushing technological development ahead. Having a new technology is good, but it is better still to find practical applications for it and figure out a way to market it.

---

[3] Article from *CERN Bulletin*, http://cds.cern.ch/record/1690337?ln=en.

[4] The 12 founding states of CERN were Belgium, Denmark, France, Germany, Greece, Italy, the Netherlands, Norway, Sweden, Switzerland, the United Kingdom and Yugoslavia.

Figure 7.7 Sumera Yamin, a physicist, and Khalid Mansoor Hassan, an electrical engineer, two scientists from the National Physics Center in Islamabad, Pakistan, during their internship at CERN. There, they learned how to make the magnets needed by the SESAME accelerator. Here they are with their completed homework, the first SESAME magnet.

Source: CERN.

Laboratories are very conscious of this and are multiplying their efforts to constantly improve. For example, at CERN, the Knowledge Transfer Office aims to list all new innovations from throughout the laboratory, and then seeks to attract business partners.

Technicians, engineers and physicists are all encouraged to notify the Knowledge Transfer Office of all new promising technological developments. This office then takes care of the task of adding value to these developments, by applying for the necessary patents and advertising them to industrial partners. Discussions are in progress with some member states of CERN to set up "incubation centers." Two such "idea nurseries" were recently established in the United Kingdom and the Netherlands. Another one will open soon in Greece.

The objective is to facilitate access to the know-how and technologies developed at CERN for small companies in high-tech sectors and to bridge the gap between fundamental science and industry. The number of potential partnerships and signed agreements about technology transfer is increasing. These efforts are contributing to reducing the time between when a discovery is made and when an application is found. In the last century, one often had to wait for several decades to see the returns of basic science become reality.

A little help can go a long way

CERN reinvests part of its benefits from knowledge transfer in new projects to support the most promising technologies, thanks to its Knowledge Transfer Fund. This is how, a few years ago, a former CERN employee received financial support from this fund to produce better-performing solar panels, taking advantage of an ultraclean vacuum technique developed for the LHC beams.

Thermal solar panels, designed to heat water using solar energy, are much less effective when the pipes carrying the hot water are not perfectly insulated. One way to improve this insulation is to place the pipes in a vacuum. This provides insulation similar to a thermos flask. An innovative method had been developed to achieve a near-perfect vacuum in the pipes carrying the beams at the LHC, to provide the best possible insulation. Here, a special material traps the air molecules left behind by the vacuum pumps. This material works more or less like the old-fashioned sticky flypapers used to catch flies. The remaining gas molecules stick to this material, yielding a near-perfect vacuum, thus eliminating heat losses.

These solar panels, which are more effective than conventional ones, now cover the roof of the main terminal at Geneva Airport (Figure 7.8). They ensure that the heating and air conditioning work, and perform well even under snow or during overcast weather.

Applications in medicine and elsewhere

Medicine is where the impact of particle physics has been most impressive. Methods used in both particle detectors and accelerators (see the box "A new weapon against cancer") have been put to use. Medical imaging has particularly benefited from physics research. This field, which started with X-rays for radiography, today offers a vast array of medical imaging techniques, such as the use of scanners, MRI and PET scans. All

Figure 7.8 More effective solar panels have been designed thanks to the ultraclean vacuum technology developed for the LHC. Some of these solar panels now provide heating and air conditioning for Geneva Airport. They perform well even in overcast conditions or under a thick layer of snow.
Source: CERN.

of these applications come directly from the work performed by physicists over the last century on X-rays, antimatter, the electron and its spin, and electromagnetism, for example. Who could have imagined any of these applications at the time?

In addition, radioisotopes, that is, radioactive atomic nuclei, are also used in medicine for both the diagnosis and the treatment of some cancers, such as cancer of the thyroid gland. Inaugurated in 2014, the CERN-MEDICIS center, a brand-new physics laboratory dedicated to life sciences and medicine, will develop new radioisotopes for medical applications.

One major problem with radioisotopes is that, since they disintegrate quickly, their use is limited to health centers located near the laboratories that produce them. CERN is working in partnership with CIEMAT (a research center dedicated to energy, the environment and technology) in Spain to develop very small accelerators. Every hospital could then be equipped with such a device and produce small doses of radioisotopes as needed, on the premises.

New weapons against cancer
....

Approximately 10,000 particle accelerators are in service worldwide for medical purposes. They can concentrate enormous quantities of energy in a microscopic point in space. Their extreme precision makes them ideal tools to target and kill cancer cells.

In addition, new types of accelerators have been designed to allow *hadron therapy*, the latest technology enlisted to fight cancer. Beams of hadrons (the name for all particles made of quarks, such as protons) are used to irradiate the affected organ, instead of the photons from X-rays used in conventional radiotherapy. Hadron therapy has the huge advantage of destroying cancer cells more effectively while not affecting healthy tissues in the process (Figure 7.9). Several hadron therapy centers are now in operation in various places around the world. Bob Wilson, the first Director of Fermilab in the US, first suggested proton therapy in 1946, and

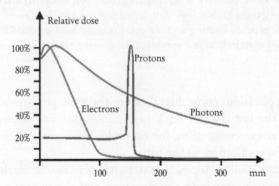

Figure 7.9 The vertical axis shows the percentage of energy deposited by various particles as a function of their penetration depth (given in mm) in human tissue on the horizontal axis. The photons from X-rays used in conventional radiotherapy have the disadvantage of losing a large fraction of their energy along the way, as indicated by the curve labeled "Photons." They thus damage healthy tissues before reaching their target, a cancerous tumor located at a certain depth. On the other hand, the protons used in hadron therapy have the immense advantage of depositing practically all their energy at a very precise depth. They can be tuned to destroy cancer cells without damaging healthy cells along the way. Electrons lose practically all their energy right at the tissue surface and are thus ineffective in destroying a tumor situated within an organ.

Source: Jean-François Héron.

New weapons against cancer (*continued*)

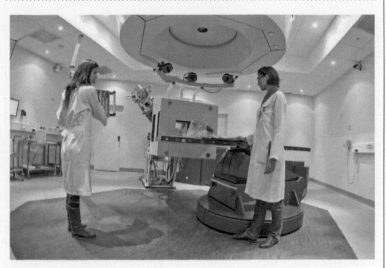

Figure 7.10 What patients see when they receive a hadron therapy treatment at CNAO in Italy.
Source: CNAO.

treatments started in the 1950s. The neutron therapy facility at Fermilab has been treating patients since 1976.

The accelerators for two such centers, CNAO (Centro Nazionale d'Adroterapia Oncologica) in Italy and MedAustron in Austria, were developed in partnership with CERN (Figures 7.10 and 7.11). A specific research program is still ongoing at CERN to improve and simplify the technology required for these accelerators. Several young researchers were also trained at CERN and now work in health centers in various countries.

Even research on antimatter has made a contribution. The ACE experiment conducted at the CERN "antimatter factory" established that antiprotons could even be more effective than protons in destroying tumors. Not only do they deposit most of their energy at a precise depth in tissue, just like protons, but antiprotons can also annihilate with protons in the atoms of cancer cells. More energy is released in the tumor, which allows even more cancer cells to be destroyed.

continued

New weapons against cancer (*continued*)

Figure 7.11 . . . and what is hidden from patients, behind the wall. This accelerator was developed in association with CERN to destroy cancer cells more effectively through hadron therapy.
Source: CNAO

The entire field of electronics and telecommunications has flowed directly from research conducted on the electron and on electromagnetic waves. Thanks to this work, we have not only radio and television but also mobile phones, positioning systems (GPS and others) and communication via satellites, as well as lasers and digital cameras. At the heart of modern computers is the CPU (central processing unit), a minuscule chip containing millions of transistors. Miniaturized printed circuit techniques have brought us a long way from the first transistor that appeared in 1947 with its very impressive dimensions (Figure 7.12).

Looking toward the near future, engineers at CERN are currently testing superconducting cables[5] that could operate at "higher temperature." We are talking here of something around −420 °F (−250 °C), a warm temperature for superconductors, which usually operate closer

---

[5] For more details, see http://cds.cern.ch/record/1693853?ln=en.

Figure 7.12 A replica of the first transistor that came out of Bell Laboratories in 1947. Today, the CPU of a computer contains millions of miniaturized transistors.

Source: Wikipedia.

to −450 °F (−270 °C), but still rather cold, even by Canadian standards. These tests are aimed at establishing the feasibility of transporting electricity over long distances without energy loss by using superconducting cables. This is a major source of energy loss in conventional power lines. Efforts are also in progress at CERN and elsewhere, such as in Belgium (see the box "Clean and safe nuclear energy"), to transmute highly radioactive nuclear waste into less harmful materials.

So, what then does basic research in physics put on our plates? Nothing too edible, granted, but the returns are still enormous. Fundamental research continues to have a major societal impact and is constantly modifying the way we live and think.

Clean and safe nuclear energy

Research in physics lies behind the development of various sources of energy. This is the case with electricity, whether it comes from solar, hydroelectric or nuclear sources. Nuclear power plants operate on the basis of *nuclear fission*, a process by which heavy atomic nuclei are broken into two or several smaller nuclei, releasing large amounts of binding energy.

Unfortunately, this technology is highly risky: it burdens the environment and future generations with radioactive waste that nobody really knows

continued

Clean and safe nuclear energy (*continued*)
. . . .

how to dispose of properly. Moreover, the nuclear reaction can turn dangerous if the control systems break down. Then, when a problem occurs, things end in catastrophe, as the accidents at Chernobyl and Fukushima have clearly demonstrated. These are not the kinds of returns I can be proud of as a physicist, even though every method of power production entails some risk: how many people have died extracting coal and oil? It is thus important to explore safer alternatives.

Research is ongoing to develop another type of nuclear reaction, called *nuclear fusion*, where very light nuclei, such as hydrogen, are merged together to form heavier nuclei. This is exactly how the Sun produces its energy. Nuclear fusion would lead to the production of an even bigger amount of energy without the dangers inherent in the classic nuclear fission power plants. Unfortunately, this technology is extremely complex and will not be completely environmentally safe, since some of the materials used will become radioactive. The international community is nevertheless investing large amounts of effort in a huge project, called ITER, located in Cadarache in the south of France.

A third, more promising alternative exists. This would allow nuclear energy to be produced in a safer and cleaner way, without generating radioactive waste, by using a particle accelerator to induce nuclear fission in a controlled manner. A variant of this technique, called ADS (Accelerator Driven System), has been proposed by several physicists, including Carlo Rubbia, the winner of the Nobel Prize in Physics in 1984 and former Director General of CERN. This consists in provoking the fission of nonradioactive atomic nuclei by bombarding them with neutrons. These neutrons are obtained by aiming a proton beam at a target of mercury, lead or bismuth. The reaction is thus controlled.

In contrast to conventional nuclear power plants, ADS reactors do not run the risk of getting out of control, since they use much less fuel and depend on an external source of neutrons to maintain the nuclear reaction. The technique is thus completely controllable, and the reaction can be stopped at will in the case of an incident or natural disaster. One can also regulate the production of energy to suit the needs of consumers, instead of constantly generating large quantities of electricity for the sole purpose of supplying enough at peak hours. Furthermore, this technology could be used to neutralize the vast majority of existing radioactive waste by irradiating it to transmute it into more manageable materials.

Clean and safe nuclear energy (*continued*)

Unfortunately, the current nuclear industry lobby is stalling (not to say blocking) the development of this technique. These people are refusing, simply because of economic considerations, to change their course, even though several experiments conducted at CERN and elsewhere have demonstrated the feasibility of the ADS technique (Figure 7.13). Nevertheless, several scientists are persisting in their efforts and are trying to rally political and industrial support. Several hundred scientists interested in this technology attended a conference held in Geneva in November 2013. Many are hopeful that an international co-operation project, called MYRRHA, will soon start in Mol in northern Belgium. This project will develop an ADS system that will burn radioactive waste from existing nuclear power plants as fuel. In the longer term, the group will develop a new type of ADS reactor that should be clean and safe.

This technique is attracting increasing interest from rapidly growing countries such as China and India, since their energy needs are enormous. Let us hope that all these efforts will soon yield a safe, environmentally friendly source of energy.

Figure 7.13 Tests are ongoing at SCK-CEN, a nuclear energy study center in Belgium, to develop a new type of nuclear reactor that would use radioactive waste from existing nuclear power plants as fuel to produce energy, and neutralize this waste.

Source: SKN-CEN with permission.

THE MAIN TAKE-HOME MESSAGE

Fundamental research in physics not only impacts on society at an economic level, but has also deeply modified our daily life. Even though not all scientific discoveries find immediate applications, without research, today we would be without medical imaging, the World Wide Web, electronics, computers and mobile phones. The branch of industry based on physics and technology generates 15% of total income in Europe and employs 13% of its workforce.

Research in physics allows us not only to increase our knowledge and answer some of the profound questions human beings have always asked about their origin and fate, but also to train a highly specialized workforce and stimulate technological development. CERN in particular is also an example of international collaboration and contributes to world peace by bringing together thousands of scientists, of 101 different nationalities, to work toward a common goal.

8

A Unique Management Model

As we saw in the previous chapter, 10,400 physicists and engineers participate in the research program at CERN, the European laboratory for particle physics. These people, called *users*, do not work directly for CERN but instead work for hundreds of institutes in 67 countries in Europe, North and South America, Asia, Africa and Australia. Today, only international cooperation can ensure the success of scientific projects on the scale of those undertaken at CERN and elsewhere in the field of particle physics.

The laboratory employs approximately 2500 other people directly (Figure 8.1), mainly scientific and technical staff recruited from member states. Fewer than a hundred physicists employed by CERN participate in basic research; the vast majority work in applied research. The laboratory takes care of all administrative and technical aspects and is entirely responsible for the accelerators, such as the Large Hadron Collider. The LHC was designed and built under CERN supervision in collaboration with industry and other laboratories, such as Fermilab in the USA and KEK in Japan. CERN staff now operate it.

On the other hand, the physics experiments are entirely the responsibility of broad international collaborations. These collaborations consist of groups of researchers employed by hundreds of institutes, operating in a vast nonhierarchical community. Each institute appoints a representative to the Collaboration Board, which determines the group's rules of operation, accepts new institutes and ensures their continued engagement. Each collaboration collectively develops its scientific project, which must then be approved by a scientific review committee appointed by the CERN Council.

As we have seen, CERN has four large detectors operating at the LHC, namely the ALICE, ATLAS, CMS and LHCb detectors. Who designed them? Who imagined, planned, oversaw and managed the work of the

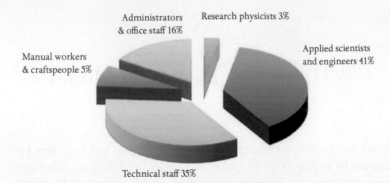

Administrators & office staff 16%

Research physicists 3%

Manual workers & craftspeople 5%

Applied scientists and engineers 41%

Technical staff 35%

Figure 8.1 The distribution of CERN staff. CERN assumes responsibility for all administrative and technical aspects of the laboratory, as well as for the accelerators. The experiments themselves are under the control of large international collaborations that attract 10,400 people to CERN.
Source: CERN.

thousands of scientists involved in these experiments? No one in particular but, rather, everyone. In fact, the scientists work practically as they see fit and as they please. Chaotic? A little, of course, but it is very effective in the end, and this is probably the only possible approach to ensure the success of such large-scale projects.

At first, no one knows the exact form the project will take. Each person involved has more or less their own ideas, which they have to debate with the whole group. Ideas and opinions evolve through discussions based on test results obtained from prototypes developed along the way. The evaluation criteria are objective: one must choose the best technology in terms of performance, reliability and cost. It would be absolutely impossible for one single person to succeed in designing and building any of these detectors.

In fact, no individual knows exactly how each detector works in complete detail. This knowledge is distributed among the whole group of scientists involved, as is the case for any large-scale industrial project. What really distinguishes these collaborations from commercial companies is that no individual dictates to another person what she or he should do. Each person and each institute must find where and how they can contribute to the success of the different ongoing research projects.

So, how does all this work? How, for example, could the ATLAS detector (one of the biggest and most complex scientific instruments ever made) be built by more than 3000 physicists and engineers from 175 institutes based in 38 different countries (Figure 8.2)? The same question goes for the CMS detector, where just as many people collaborate, or for the thousand scientists involved in ALICE and LHCb. What motivates all these people? How do they manage to work together and push back the limits of what is feasible?

A single common objective

The cohesion of these collaborations depends on the existence of a common objective: to understand what the fundamental components of matter are and how these particles interact with each other. These scientists are trying to understand the Universe, how it was formed and where it is heading. This is a major challenge that only a highly motivated team can accomplish. The motivation comes from basic scientific curiosity, the insatiable need human beings have to understand the material world we inhabit. This curiosity is essentially the same as that which motivates you to read this book, for the sole purpose of understanding a little more. This shared motivation determines the direction of the work of each collaboration.

The scientists have to work out a strategy to allow them to answer some of the big questions of their time. When the four LHC collaborations were formed in the early 1990s, one goal that the physicists from CMS and ATLAS wanted to accomplish was to establish or reject the existence of the Higgs boson. But this was only one of the many hypotheses and unanswered questions the LHC scientists wanted to address.

The main objective of the LHCb experiment is to understand where all the antimatter that was produced after the Big Bang has gone. The ALICE Collaboration (Figure 8.3) wants to determine how matter formed after the Big Bang. Everyone wonders about the nature of dark matter. What is the "new physics" that would explain phenomena beyond the reach of the Standard Model? Will supersymmetry turn out to be the right answer?

Distribution of All CERN Users by Nationality on 24 January 2018

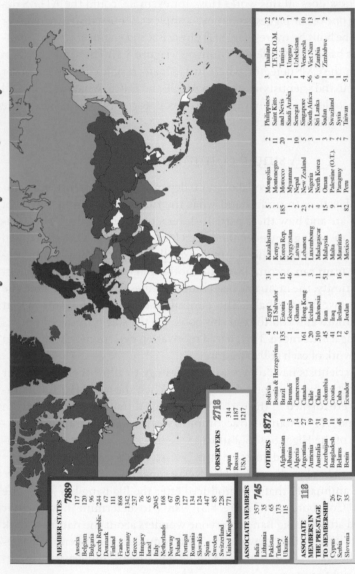

MEMBER STATES **7889**

Austria	117
Belgium	120
Bulgaria	96
Czech Republic	244
Denmark	67
Finland	111
France	868
Germany	1342
Greece	237
Hungary	76
Israel	65
Italy	2045
Netherlands	168
Norway	67
Poland	350
Portugal	127
Romania	134
Slovakia	124
Spain	447
Sweden	85
Switzerland	228
United Kingdom	771

ASSOCIATE MEMBERS **745**

India	357
Lithuania	35
Pakistan	65
Turkey	173
Ukraine	115

ASSOCIATE MEMBERS IN THE PRE-STAGE TO MEMBERSHIP **118**

Cyprus	26
Serbia	57
Slovenia	35

OBSERVERS **2718**

Japan	314
Russia	1187
USA	1217

OTHERS **1872**

Afghanistan	1	Bolivia	4	Egypt	31	Kazakhstan	5	Mongolia	2	Philippines	3	Thailand	22
Albania	3	Bosnia & Herzegovina	2	El Salvador	2	Kenya	3	Montenegro	11	Saint Kitts and Nevis	1	T.F.Y.R.O.M.	2
Algeria	14	Brazil	135	Estonia	15	Korea Rep.	185	Morocco	20	Saudi Arabia	1	Tunisia	5
Argentina	27	Burundi	1	Georgia	46	Kyrgyzstan	1	Myanmar	1	Senegal	10	Uruguay	1
Armenia	19	Cameroon	1	Ghana	1	Latvia	2	Nepal	2	Singapore	5	Uzbekistan	4
Australia	31	Canada	161	Hong Kong	3	Lebanon	23	New Zealand	4	South Africa	3	Venezuela	10
Azerbaijan	10	Chile	20	Iceland	3	Luxembourg	2	Nigeria	2	Sri Lanka	6	Viet Nam	13
Bangladesh	11	China	510	Indonesia	11	Madagascar	4	North Korea	1	Sudan	3	Zambia	1
Belarus	48	Colombia	10	Iran	51	Malaysia	15	Oman	3	Swaziland	1	Zimbabwe	2
Benin	1	Croatia	11	Iraq	41	Malta	9	Palestine (O.T.)	7	Syria	2	
		Cuba	12	Ireland	16	Mauritius	1	Paraguay	2	Taiwan	51	
		Ecuador	1	Jordan	6	Mexico	82	Peru	7			

Figure 8.2 The 101 nationalities and distribution of the 10,400 CERN users as of January 2018

Source: CERN.

Figure 8.3 Part of the ALICE detector, which is dedicated to studying the behavior of matter in the first moments that followed the Big Bang.
Source: CERN.

Getting the tools needed

All of these questions pushed experimentalists to imagine a huge particle accelerator, the Large Hadron Collider, as well as its four gigantic detectors (ALICE, LHCb, CMS and ATLAS) (Figure 8.4). The idea got around in the particle physicists community, and people interested in these projects started meeting on a regular basis to determine together the characteristics of the tools they needed to test the maximum possible number of hypotheses. The objective was to answer as many of these questions as possible. And that was how the LHC was born.

The role of experimentalists is to test the most plausible theoretical hypotheses. Theorists, on the other hand, lean on what is already established, that is, everything that has been revealed by experiments in the preceding decades, and develop new theories to achieve a better description of the physical world that surrounds us. These scientists also have to imagine the behavior of the new particles that come with their hypotheses. For example, they had to predict in advance, long before it was discovered, how the Higgs boson might be produced and how it

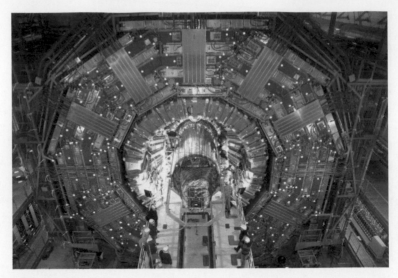

Figure 8.4 Insertion of the biggest silicon tracking device in the world into the CMS detector.
Source: CERN.

might decay. These hypotheses were then used by experimentalists and guided them in deciding on the best strategy and the best possible tools to discover new particles or test various hypotheses.

To achieve this goal, physicists need two main tools, as we saw in Chapter 3: an accelerator to produce new particles, and a detector to detect them. The latter is nothing but a gigantic camera that captures images of the new particles produced through their decay products.

Starting from the main objective

The accelerator was thus designed to be as powerful as possible to maximize the chance of producing particles that had never been made in a laboratory before. CERN already had a 27 km tunnel that housed the predecessor of the LHC, LEP (Large Electron Positron collider). The same tunnel was reused to reduce the construction costs, although it needed to be entirely refurbished with superconducting magnets, which are much more powerful than conventional magnets. Instead of accelerating electrons, the accelerator is now powerful enough to bend the trajectories of protons, particles that are 1836 times heavier than electrons. In doing so, the collision energy went from 200 GeV in LEP to 13,000 GeV (or 13 TeV) in the LHC.

Once the accelerator parameters were known (the energy of the collisions, their frequency and the number of collisions expected per second), scientists could define the characteristics of the detectors to maximize the chances of new discoveries, not only of the Higgs boson but also of a whole zoo of hypothetical particles with all sorts of characteristics. While the search for the Higgs boson was by far the most prominently featured activity in the media, the quest for supersymmetry, dark matter and the first signs of new physics has been on the research agenda for the experiments since the very beginning.

The scientists had to design detectors that were as versatile as possible, since nobody knew exactly at the time how the Higgs boson and all the other hypothetical particles that were being sought were going to manifest themselves. The Monte Carlo simulations described in Chapter 4, based on theories developed by theorists, guided their choices of the properties of the detectors.

Getting the tools needed

How do we design our detectors? The starting point is our common goal: to push human knowledge a notch further by finding new particles and measuring their properties to test various hypotheses. Each of the four LHC experiments was formed around a common project—the specific questions that they aimed to resolve—and this objective guided them in designing the best possible detector.

In Chapter 3, we saw that a detector is used to reconstruct events, that is, to determine from their decay products what particles are produced during the proton collisions generated by the accelerator, the LHC. The detector must have the capability of identifying all types of lighter and more stable particles that might emerge from these decays.

A detector is made of several layers, like an onion, each layer corresponding to a different subdetector that is aimed at extracting part of the information about every single particle crossing the detector, as is shown in Figures 8.3–8.7 for the four detectors operating at the LHC. We need to reconstruct their trajectory, estimate their energy, determine their electric charge and figure out their identity. One or several subdetectors are needed to measure each of these properties. Each

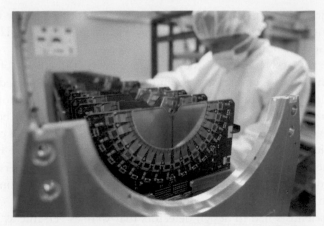

Figure 8.5 Assembly of the 42nd and last module of the vertex locator (VELO) of the LHCb detector in a cleanroom. This detector can determine the origin of every electrically charged particle, hence allowing reconstruction of the particles initially produced.

Source: CERN.

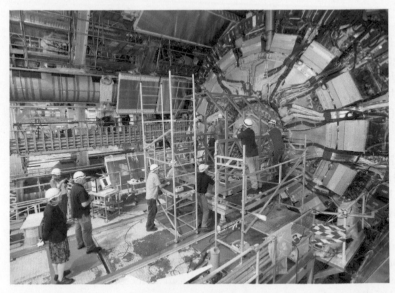

Figure 8.6 Insertion of the central part of two tracking devices into the heart of ATLAS.

Source: ATLAS.

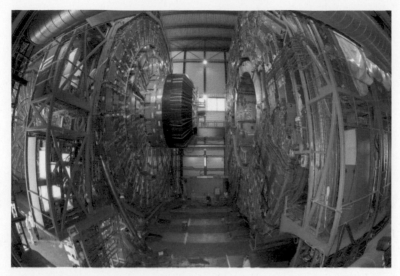

Figure 8.7 The CMS detector before final closure. Cosmic rays were used to conduct the first tests involving all subdetectors as a final whole-detector check before the LHC was turned on.
Source: CERN.

subdetector and every other tool needed for these tasks becomes a project. Within each collaboration, scientists assign themselves to a specific project according to the needs of the experiment but also according to their interests, their resources and their expertise, so that the scientists can guarantee that they will complete these projects. Everyone must also contribute to common tasks, such as data taking in the control room as illustrated in Figure 8.8.

A huge scientific picnic

The collaborations have established some rather flexible rules, although these lack any legal status, that simply stipulate that each participating institute must contribute to one or another of these projects. Such a contribution might be the design or fabrication of a subsystem used to reconstruct particle tracks. It could also involve the development of algorithms and software dedicated to analyzing the data. Other institutes might have to supply a data distribution system to ensure the reconstruction of events on thousands of computers operating in parallel.

Figure 8.8 Kerstin Jon-And, former head of the ATLAS Collaboration Board, taking a shift in the control room. This board oversees the elections to key coordination positions and determines various policies for the collaboration. Everybody participates in data taking in the control room, except for rare exceptions made for people occupying the most demanding posts.
Source: CERN.

This works more or less like the basic principle of communism, "From each according to their ability." Each institute contributes according to its resources, which depend generally on what the funding agencies from their country can afford. Different institutes from the same country also have to agree among themselves about how to share the resources assigned to them. They also have to hire the staff needed for the completion of the tasks that they have agreed to do. Large scientific collaborations such as those at CERN follow a mode of operation completely different from what is common in large companies or other international organizations. It looks pretty much like a huge "scientific picnic," where each participating institute agrees to contribute something. As in a collective meal, each group brings what it wants, although someone coordinates the whole to make sure that there will be enough to eat and drink.

For the LHC collaborations, all items required for the detector are described in a technical document that is prepared beforehand and approved by the whole collaboration. Each institute decides freely on its contribution but has to demonstrate to the whole collaboration that it will be able to assume this responsibility. The final distribution of tasks is agreed upon by consensus. This is a dynamic process that evolves throughout the project.

Once the tasks have been distributed and the schedule established, the project coordinators follow closely the progress of all tasks placed under their responsibility. Every group or institute must complete its part of the project so that the "picnic" is a success. But the responsibility remains collective: if a group assigns a task to an institute and that institute then struggles to achieve its goal, the whole group looks for means (financial, technical or human) to support the institute that is having difficulties. In the end, nobody can succeed and achieve the shared goal if any subsystem fails.

Who does what?

Who does what, then? And who decides? Nobody; there really is no central authority. The approach relies on the whole group, not on following orders from above. The project draws on the creativity and expertise of everyone involved. Nobody dictates his or her vision of things, although several people would love to do so. Of course, since we are dealing with human beings, some people's egos will be offended at times in the process. Nevertheless, all participants must discuss their ideas and convince the others, and then a consensus must be reached. When different approaches are proposed within a working team, each person or each working unit has to demonstrate the merits of their ideas to the whole group, by using either simulations or results from tests done on prototypes. This occurs during countless (and sometimes endless) meetings held at various levels. Just for fun, I counted 75 working meetings for the ATLAS Collaboration the day I wrote this. These meetings may be held at CERN or elsewhere, and most are accessible via videoconferences. This way, even researchers based in their home institute can participate in these meetings.

Presenting one's ideas in front of a group often brings possible improvements to light or reveals weaknesses. In the end, decisions are taken collectively and are based on scientific evidence demonstrating the benefits of a particular approach. When there is a divergence of opinion, the scientific interest has to prevail above all else and guide all decisions. It is thus the common scientific objective that determines the functioning of the whole collaboration. And all decisions are taken by consensus.

Here is an example to illustrate this. During the construction of one of the ATLAS tracking systems, a major problem was discovered halfway through the project. It became obvious that we had to either redo a substantial part of the construction, or change the gas planned for that subdetector. Both solutions entailed many risks and inconveniences. After several days of debate, where everyone was able to state her or his opinion and present the results of the various tests they had done, the whole group, that is, about 50 people minus one, agreed to develop a new gas mixture for the detector. The only dissident was the project coordinator himself. Since he had failed to convince the group of the superiority of his opinion, he had to defer to the majority.

In a commercial project, this man would probably have been able to impose his view, at the risk of course of losing his job later on if he had made the wrong choice. But in an LHC collaboration, the whole group assumes responsibility for the project. It is a collective effort and everyone shares in its success, as seen in Figure 8.9.

Figure 8.9 The LHCb control room just after the first beams collided on March 30, 2010 at the restart of operations following the major technical incident in 2008.

Source: CERN.

The ideas always progress, but rarely in a straight line. New technologies usually need to be developed along the way to achieve the goals that have been set. Furthermore, we always have to balance the needs of the research objectives on the one hand against what is feasible on the other hand, and so the detector characteristics must be adjusted accordingly. Once the scientific criteria have been determined, the physicists hand their projects over to engineers. Several details may still have to be adjusted along the way to deal with unforeseen difficulties.

So nobody ever woke up one morning thinking that the LHC and its four detectors must be built. Instead, all of this took shape over a period of about 15 years, during which the details of the project were constantly questioned and revised. In these large scientific collaborations, there are no directors, only coordinators. Every decision is made by consensus. The debates are generally beneficial, even if human nature sometimes complicates the discussions. But in the end, the common scientific objectives always prevail and have the last word.

Motivation and tolerance

One must bet on the personal motivation of all those involved to ensure that every single aspect of the project is realized in time and within the best possible criteria. Fortunately, scientific curiosity drives all of the participants and supplies the motivation and commitment needed for the project to succeed. And, although scientific curiosity is the main motivation, the hope of increasing one's status within the team also plays a prominent role. Work within multicultural and international teams, on projects at the forefront of innovation, and where one is free to do as one pleases, brings great pleasure (see the box "A small world"). But there is never any financial compensation for achieving an objective.

To succeed, we need a lot of tolerance and we must value diversity, cultural as well as individual, as we shall see in the next chapter. This is what makes it possible for people of so many different nationalities to work together. For example, at one point, there was a team of technicians from Russia, Israel, Pakistan, the United States, China and Japan. Together, they installed the giant muon wheels of the ATLAS detector (Figure 8.10) under the supervision of a French engineer. All of this was accomplished while communicating in English of widely variable quality.

Figure 8.10 Members of technical teams from Pakistan and Israel pose in front of part of the big muon wheel of ATLAS during its construction. People from these two groups became friends and took advantage of their stay in Europe to go sightseeing together, such as spending a few days together visiting Paris. *Source*: CERN.

A small world

Working in particle physics certainly gives one opportunities to participate in outstanding projects. But for me, the most enriching part has been to do so within such diversified teams. During the 19 years I spent at CERN and the 5 years I worked at SLAC and Fermilab, two laboratories located in the United States, I sat alongside people from dozens of different countries and had exchanges and discussions with them, laughed with them and ate with them. My last working team contained a dozen people, coming from India, Pakistan, the United States, the Netherlands and Canada.

Over the years, I have worked directly with one Chinese person, one Australian, Russians, Greeks, Turks, Swedes, one Taiwanese, one Togolese, Koreans, Algerians, one Colombian, Americans, Spaniards, Serbs, Germans, Indians, one Vietnamese, Japanese, Brazilians . . . OK, I'll stop, the list is much too long. There were always so many people of different nationalities that I often counted them during meetings just for fun. I was able to

A small world (*continued*)

exchange comments with people from everywhere in the world and learn how they worked and approached things. I now have friends from almost everywhere on the planet with whom I can discuss any subject. It changed not only the way I cook, but also and especially my vision of the world and my understanding of the political situation of countries and their history. There's nothing like a discussion with people from a foreign country to understand better its politics, history or geography.

I had countless discussions with women and men about the situation of women in their country. I was able to talk with people who had known war, poverty and natural disasters that I had only briefly heard about on television. Knowing so many different people allowed me to discover other cultures and to be more open-minded.

The most surprising thing in the end is to realize how similar we all are, in spite of cultural differences. This is so true that it is extremely easy to forget when working in such a community that the person in front of you comes

Figure 8.11 24 hours a day, teams of physicists take turns in control rooms to ensure that the data collected is of good quality. This is always an excellent place to meet colleagues of diverse origin.
Source: CERN.

continued

A small world (*continued*)

....

from a country located thousands of kilometers away from yours, with a completely different culture, language or religion (Figure 8.11). Everybody eventually learns to use the same language, a mixture of technical and scientific terms delivered in English of various flavors. Everybody shares the same purpose and the same passion. This is what allows us to easily overcome any difficulties or fear generated by difference.

To make this work, everybody must be willing to lend a hand, no matter how menial the task may seem. For example, one physicist spent months installing cables on a detector with a technical team. Why? Simply because she wanted to be certain that everything would work perfectly and she would have the immense pleasure of reaching her scientific goals. Her case is far from being unique. I could cite practically every one of my 3000 colleagues. I myself spent 2 years with an engineer, a few technicians and a couple of students testing every single one of 118,224 wires in the central part of one of the tracking systems.

A life dedicated to research

It is not rare to see physicists dedicating a substantial part of their career to the pursuit of one precisely defined objective, be it the discovery of the Higgs boson or supersymmetric particles, or solving the dark matter mystery. When someone has dedicated one or more decades to working on a problem, they will not hesitate to spend a year or two on checking hundreds of thousands of small wires in a subdetector to make sure that it will run smoothly. We know that it is essential if we want to achieve our final goal.

Every person thus contributes to very diverse tasks, following the needs of the experiment. During the design phase of all the LHC detectors, countless feasibility tests were conducted with prototypes. When we moved on to the subdetector construction phase, we constantly checked the quality of all components. The assembly of the detectors underground also required the acquisition of new skills. Several people had to be trained to work on scaffolding or to use rock-climbing equipment. Finally, now that the operation phase has begun, all of the scientists involved take turns in the control rooms to make sure that the data is of good quality. Teams are formed to intervene quickly in cases of equipment failure or software problems. In connection with the data analysis, physicists meet regularly to discuss their results, improve the calibration, refine the analysis methods and develop better software.

In the end, each of the LHC detectors is an incredible scientific instrument that combines gigantism and high precision. For example, the ATLAS detector weighs 7000 tons, the sum of millions of small, handmade, delicate components. It is crisscrossed by 3000 km of cables of all kinds that feed the beast with high and low voltages and collect the signals coming out of its 100 million channels. There are just as many pipes and pieces of tubing of all sorts carrying cooling liquids and various gases everywhere. It is a miracle that all of this works! (Figure 8.12)

But there is a flipside to the coin. Although it is not a characteristic unique to scientists, this passion for research often turns into near-obsession and work bulimia. Several people work all hours of the day and night, neglect their family or their health, organize meetings on Saturdays, do not take holidays, or fire off replies to e-mails faster than

Figure 8.12 Everybody contributes to the project. Given the complexity of these instruments, it is a small miracle that they work. Here is part of the ATLAS detector during its construction.

Source: CERN.

their own shadow, no matter what the time of day or night. In other words, they sacrifice everything for their work. Personally, I do not think this is necessary, certainly not in continuous mode, although the high-responsibility positions often require it.

A democratic model

There is neither a director nor any managers to oversee everyone's work, only a spokesperson whose role is to overview the operation. Delegates from each participating institute elect coordinators for the various projects. A person is sometimes elected to a post partly for political reasons. We try as much as possible to ensure a degree of diversity in the positions at coordinator level. This helps to ensure that people from countries with fewer resources who cannot spend as much time based at CERN, where it is easier to take part in teamwork and represent one's ideas better, are not disadvantaged. These coordination posts are very demanding but also very stimulating. Having more responsibility allows individuals to develop professionally, show their full potential and move upward.

Collaboration and competition

Researchers have to find the right balance between competition and collaboration. The tacit and most important rule is that one must collaborate. The working groups eventually push aside those who refuse to do so. In any case, no one can do everything on their own; everyone needs the others, and solid common tools. Knowledge and resources must be shared.

Nevertheless, every collaboration member is in competition with the others. The vast majority of researchers, especially the younger people, have only short-term contracts. Each person must therefore show their ability and demonstrate their value if they want to obtain a permanent post in one of the participating institutes. It is not easy to secure a position and be given the opportunity to continue taking part in this extraordinary scientific adventure. Given the rarity of permanent positions, even talented people do not always manage to stay in the field.

Individual contributions are known and recognized at the level of the working groups, even though everything is achieved through collaborative effort. In the end, no one can claim all the credit. The work of everyone is taken into account. To acknowledge this, every scientific publication is signed by every scientist of the collaboration, that is, approximately 3000 people in the case of CMS and ATLAS.

Of course, generally, only a dozen or so people will have performed the specific physics analysis leading to the publication of an article in a scientific journal. But without the contributions of all who participated in the design, construction, installation, calibration and operation of the detector, not to mention simulations, software development and computing management, no publication would be possible. Every person whose name appears on one of these publications can say proudly say that he or she played a role in this adventure. Key people, whose contributions were particularly important, are chosen to present the collaboration's scientific results at the most prestigious international conferences.

These collaborations function well because all of the people involved are eager to see their experiment succeed. There is no coercion or financial bonus. The peer recognition and the huge satisfaction that they experience from contributing to the success of such unique projects, aimed at pushing the limits of knowledge, is the main motivation for the vast majority of participants.

THE MAIN TAKE-HOME MESSAGE

The experimental collaborations at CERN, such as ATLAS and CMS, each have more than 3000 researchers, coming from the five continents. Nobody manages the group centrally in a command–control fashion. Instead, each person is expected to fit in and contribute to the best of her or his ability to ensure the success of the experiment. This enterprise works essentially because of the existence of a common objective driven by scientific curiosity, the commitment of each individual and the tolerance of everyone involved. It works like a huge picnic where each group contributes a dish according to its tastes, resources and talent. The desire to see the experiment succeed is what motivates everybody. All decisions are taken by consensus within a collaboration, using agreed coordination mechanisms. This model gives power to the base and draws on the talents of all members. Project coordinators make sure that all aspects of the work are covered. The process is sometimes a little chaotic, but it is necessary in order to let creativity flow freely and set the stage for revolutionary discoveries. In the end, this may be the only possible way to complete such large-scale projects.

9

Diversity in Physics

Creativity is essential to the scientific process and paves the way for discoveries. As we saw in the previous chapter, large collaborations in particle physics rely on the exchange of ideas through discussion to determine the best possible strategy, from detector construction to data analysis. Creativity draws its inspiration from diversity and thrives on a variety of opinions. The more numerous the different approaches, the better the ideas that come out.

Nevertheless, anyone visiting an international research laboratory such as CERN will notice the overwhelming proportion of men and the dominance of people of Caucasian origin, even though the scientists are of 101 different nationalities. Men occupy 82% of all scientific positions at CERN, and there are just as many people of Caucasian origin. The fact that CERN was originally a European laboratory only partially explains this situation.

Traditionally, the physics, mathematics and engineering communities have been rather conservative. Numerous initiatives to increase their diversity have appeared over the last few decades, aimed especially at attracting more women and also, though to a smaller extent, people of different races and ethnic groups. These initiatives are bearing fruit and, as we shall see, the proportion and visibility of minorities in particle physics are increasing, which is very encouraging.

But this battle is far from over, not only for women but also for disabled people, for people of racial or ethnic groups and religions different from the majority, and for members of the LGBT+ community (lesbian, gay, bisexual, transgender etc.). It would be great if diversity were taken into account at the time of hiring and, more importantly, if efforts were made to help ensure that mentalities evolve so as to develop a work environment where everyone feels comfortable. This is essential if one wants to retain members of minority groups (see the box "How to attract, hire and retain minorities in science"). No matter what the job, everybody prefers to work in a welcoming rather than a hostile environment.

As in Chapter 7, on the benefits of basic research, I will give examples mainly from CERN since, because of its international composition and its size, it provides an excellent overview and good statistics. To support my viewpoint, I will also draw material from various studies, including a massive survey conducted on 15,000 physicists from 130 different countries.

Female physicists at CERN

Before going any further, let's examine the situation. As explained in previous chapters, there are two categories of workers at CERN. In 2014, CERN employed 2513 people, of whom 44% were engineers and physicists, mostly working in applied research. In this category, 12.2% were women.[1]

Most of the researchers involved in fundamental research, however, belong to the second category of workers. These people called *users* were employed by hundreds of institutes from 67 different countries and participated in the research program of the laboratory. Of this group of 10,416 people, 85% were physicists and 9% were engineers, the rest consisting of technical and administrative staff.[2] The fraction of women among CERN users as of September 1, 2014 was 17.5%. This might not seem like much but it is already considerably better than it was 10 or 20 years ago, and it is continually improving (Figures 9.1). For example, in 2008, only 15.6% of the physicists in the ATLAS Collaboration were women. This fraction reached 19.6% in October 2012 but, two years later, still stood at 19.7%. The figure varies slightly between experiments, and widely from one country to the next, as can be seen from Table 9.1.

The graph in Figure 9.2 shows the distribution of CERN users by age group. The average age is at present 41 years. This corresponds to an average age of almost 42 years for men and slightly more than 37 years for women. More women are coming in, although this trend is recent—the women are younger—as indicated by the gap between the average ages for the two sexes.

---

[1] Official CERN statistics as of December 31, 2013, https://cds.cern.ch/record/1703227/files/CERN-HR-STAFF-STAT-2013.pdf
[2] Data as of September 1, 2014, provided by CERN.

Figure 9.1 Physicists from the LHCb Collaboration running the experiment on International Women's Day on March 8, 2010. Hundreds of female physicists marked the date by making themselves slightly more visible on that day, to show how much progress had been made. On that day, from 7:00 until 23:00, teams made up only of women staffed the LHC, as well as the ALICE, ATLAS, CMS and LHCb control rooms.

Source: CERN.

Table 9.1 Percentages of female scientists at CERN.

CERN users by nationality	Percentage of women	Percentage of women below 35 years of age	Percentage of people below 35 years of age	Total number of people at CERN
Turkey	33%	40%	59%	159
Norway	29%	33%	41%	59
Greece	28%	32%	38%	152
Romania	26%	30%	36%	121
Belgium	25%	25%	54%	109
Spain	25%	31%	38%	323
Sweden	24%	36%	39%	71

continued

Table 9.1   (*continued*)

CERN users by nationality	Percentage of women	Percentage of women below 35 years of age	Percentage of people below 35 years of age	Total number of people at CERN
Italy	23%	31%	29%	1666
India	23%	26%	52%	214
Bulgaria	22%	44%	22%	74
China	22%	23%	72%	302
Portugal	20%	21%	45%	104
Brazil	20%	12%	54%	111
South Korea	19%	23%	49%	115
Finland	19%	21%	30%	79
Mexico	19%	28%	58%	69
Poland	19%	16%	39%	247
France	17%	25%	26%	731
Slovakia	17%	21%	51%	102
Canada	16%	22%	48%	141
Israel	15%	29%	33%	52
United States	14%	18%	41%	973
Germany	14%	19%	47%	1095
Switzerland	14%	18%	31%	177
United Kingdom	12%	17%	46%	633
Hungary	12%	22%	34%	67
Russia	11%	18%	22%	951
Austria	11%	15%	33%	81
Netherlands	10%	28%	25%	144
Ukraine	10%	14%	58%	60
Denmark	9%	21%	36%	53
Czech Republic	9%	10%	51%	216
Japan	7%	8%	47%	253

Note: The 33 countries with more than 50 users are listed here in decreasing order based on their percentage of women scientists at CERN.

Source: Pauline Gagnon, based on CERN data as of September 1, 2014.

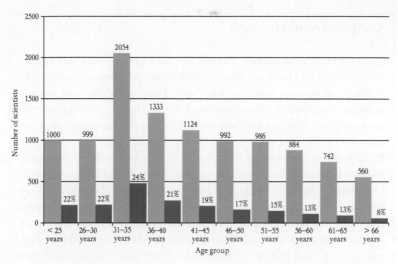

Figure 9.2 Distribution of CERN users by age group. The paler columns indicate the number of users of all genders whereas the darker columns show the percentages of women in each age group.

Source: CERN.

Analysis of results

Let us examine the situation for the proportion of women by nationality at CERN. Only countries with more than 50 users, that is, people of the same nationality conducting research at CERN, were included in Table 9.1, to ensure reliable and meaningful statistics. The complete list of the 101 nationalities represented at CERN is given in Appendix A. The countries are listed in decreasing order; that is, countries with the largest fraction of women come first.

All CERN users were classified by nationality based on their passport, regardless of their affiliation (the country of their employing institute or university). For example, I am counted in the statistics for Canada, even though I work for an American institute. Turkey is where we find the highest proportion of female physicists. The highest percentages are in the Balkans (Turkey, Greece and Bulgaria), elsewhere in Europe (Norway, Romania, Belgium, Spain, Sweden and Italy), and India. On the other hand, countries such as Japan, Austria, Switzerland, Germany, the United States,

continued

Analysis of results (*continued*)
....

Canada and the United Kingdom are below the average. What could explain these differences? The causes are diverse and there is no unique explanation, although salary level seems to play a role. There are often more women in physics in countries where salaries are lower. This suggests that women are more easily taken in for lesser paid jobs while the better paid positions are reserved for men since they are still perceived as the "breadwinner." But this is only one of many contributing factors. For example, Russia and the Czech Republic have very few women, although the salaries there are low. Other cultural and historical elements come into play as well. Every country is unique and must be analyzed individually. You may notice in the complete list given in Appendix A that some countries have a proportion of women exceeding the value of 33% found for Turkey. However, these values refer to very small groups of people and are therefore meaningless from a statistical viewpoint.

Looking at the proportion of women in the group of people below 35 years of age (the third column of figures), one can predict how the representation of women will evolve over the next 5–10 years. Nearly all of the countries listed here have a higher percentage of women in the younger age group, except for Brazil and Poland. These two countries have proportionally fewer women among the youngest members of their groups.

If the general trend continues, the percentage of women among CERN users should go from 17.5% today to approximately 19% in 5 years. Although this is positive, at this rate, equality is not for tomorrow morning . . . CERN and all countries involved in it must continue promoting diversity and multiply their efforts to attract more young women and people from other minority groups into scientific careers. Joint efforts must also be undertaken to hire and retain more minorities.

There is a slight decrease in the youngest group. Hopefully, this is just a small fluctuation and not a trend that will remain in the coming years. This decrease should serve as reminder that the battle is far from over. Several measures can be taken to fight stereotyping and continue to attract more young women and other people from underrepresented groups into science. Several suggestions are given in the box "How to attract, hire and retain minorities in science."

Why are there so few women in science?

There are multiple reasons why there are so few women in science, but one major problem is rooted in stereotypes and acquired attitudes. No one has ever been able to demonstrate the superiority of boys in science or the existence of biological differences that would justify a belief in such a superiority, quite the contrary. At high school level, several studies have in fact shown that girls succeed slightly better than boys in science and mathematics.

According to Catherine Vidal,[3] a neurobiologist at the Pasteur Institute in Paris, there is no significant difference of biological origin between the sexes in terms of capacities. This observation comes from analyzing brain activity through imaging studies. Vidal also stresses that only 10% of all synapses (the connections between neurons in the brain) exist at birth. The remaining hundreds of billions build themselves through learning. She thus uses the term "intellectual plasticity" to describe "the shaping of the brain under the influence of the environment, be it internal (food, hormones) or external (family and social interactions)." Everything is thus mostly the result of educational, culture and social pressure.

This is precisely what Annette Jarlégan, a lecturer in education sciences at the University of Nancy 2, studied. She demonstrated how all this takes place subtly without anyone noticing. For example, teaching aids, such as certain children's books, encourage boys to partake in an active and public life, whereas these books often suggest more passive roles to girls, confined to the private sphere (the home). She quotes several studies demonstrating that teachers pay more attention to boys than to girls. Their expectations are also higher for boys. The two groups do not receive the same type of encouragement. The same assignment is graded differently by a group of teachers of both sexes if this assignment bears the name of a boy or a girl. This is even more noticeable for scientific subjects. All of these small differences constitute what she calls the hidden curriculum, "a set of values, skills and knowledge that school kids acquire without being aware of it, without having it registered on the official programs, with teachers and even the parents being unaware of it."

---

[3] http://cordis.europa.eu/news/rcn/30550_en.html

"The stereotypes found between boys and girls at school are the same observed between working class and upper class," notes Jarlégan. "The girls are said to be persistent, brave, they succeed thanks to their efforts. This is exactly what is said of [kids coming from] lower classes." Deep down in their heart, boys know that the world belongs to them. They can have fun during primary school and fool around in high school and college, since jobs in the best sectors are awaiting them. The girls, on their own, will exclude themselves massively from these areas.[4]

Girls score more poorly in mathematics tests if they are reminded prior to the test that, in general, women are less successful in mathematics. This is what the researchers Steven Spencer, Claude Steele and Diane Quinn discovered and called the "stereotype threat."[5] Girls performed better in the same tests if they were told instead, before the test, that there is no known difference in the performance of the two sexes. All boys, regardless of their ethnic origin, as well as all children of Caucasian origin, performed better if their origin was underlined before the test. This effect is called the "stereotype boost." This demonstrates that social context impacts on the performance of girls and other minorities in science.

It is not easy, then, for girls to choose a scientific career when everything reminds them that they do not belong there. This message can come from their family, their school or the media. Textbooks and the media almost exclusively refer to scientists as male, and images of male scientists reinforce this message (Figure 9.3). Nobody questions the relevance of a man's presence in physics, engineering or mathematics. Nevertheless, many women are asked how come they have chosen such careers. Such remarks underline the suggestion that they do not belong there. Not surprisingly, only the best and most determined persist. Without support, some young women lose courage and opt for sectors where they do not need to constantly swim against the tide.

[4] https://antisexism.wordpress.com/2011/11/19/inequality-between-girls-and-boys-at-school/

[5] http://www.leedsbeckett.ac.uk/carnegie/learning_resources/LAW_PGCHE/SteeleandQuinnStereotypeThreat.pdf

Figure 9.3 The team of physicists running the CMS experiment on March 8, 2010.
Source: CERN.

Discrimination at the hiring stage

The Discrimination Observatory[6] in France and many other groups elsewhere in the world have conducted numerous studies showing that there are several forms of discrimination at the hiring stage. It can be based on, among other things, gender, age, appearance, disability, race or origin. Employers generally prefer to hire someone who looks like them. Furthermore, gender stereotypes continue to impact on employers, be they male or female.

A survey[7] conducted at Yale University in the United States revealed that women were just as likely as men to maintain sexist stereotypes. The same fictitious CV was sent to 127 physics professors. Each professor was asked to evaluate the CV and decide whether he or she would hire that person as a laboratory assistant. Half of the CVs distributed were

[6] http://fr.wikipedia.org/wiki/Discrimination_%C3%A0_l%27embauche
[7] http://physicsworld.com/cws/article/news/2012/oct/24/physicists-show-bias-against-female-job-applicants

signed with a man's name, "John," and the other half with a woman's name, "Jennifer." Except for the name, the CVs were identical. Nevertheless, John's CV was evaluated more positively by both male and female professors. These potential employers judged John as more competent. They even offered him a higher salary, on average $4000 more annually. Studies of this kind suggest that applications ought to be made anonymously, that is, without having the name of the candidate on the CV, to avoid this kind of discrimination. This would also benefit people whose name does not conform to the standard.

How to attract, hire and retain minorities in science

Here, in essence, are a set of recommendations that were initially formulated by a group of young women working at CERN and presented at a meeting of the Economic and Social Forum (ECOSOC) of the United Nations Organization held in March 2013. I have adapted them to apply to all minorities as well. Several of these recommendations would also contribute to improving the work environment in science and hence be of benefit to all.

To attract more members of minorities into science, we could:

- *Fight stereotypes at all levels.* We should improve the representation of minorities in textbooks, including in the phrasing of problems; use gender-neutral and multicultural language when referring to scientists; and increase the visibility of scientists from minority groups in the general culture by providing more diverse contacts for the media.
- *Help young people build a strong "physics identity."* Students who do not feel good at mathematics or science do not pursue a career in it. Encouragement from peers, teachers and family helps young people to believe in their own ability. Classroom activities such as discussions on cutting-edge physics topics, being encouraged to ask questions and peer teaching all contribute to building a strong "physics identity." Having discussions on why there are fewer women and people from other minorities in science also helps young people from minority groups see that the problem does not come from them but has social roots.
- *Provide role models and mentors for young people from minority groups.* We should give opportunities to LGBT+ people, women, and people of different races and faiths to talk about their career to young audiences. This should be

How to attract, hire and retain minorities in science (*continued*)

done at all stages. Careers fairs should be held to reinforce young people's self-esteem and provide a context where they can have discussions with other young people facing similar challenges.

To hire more members of minorities in physics and science in general, we could:

* *Implement anonymous job application processes.* The applicant's gender, race and marital status should be hidden during the job application process to avoid gender bias, since studies have revealed that both men and women discriminate against women. The number of female musicians in five major orchestras tripled once job applicants had to perform behind a curtain.
* *Implement equitable parental leave.* Both men and women should be given parental leave, and men should be strongly encouraged to take it. Young women of childbearing age would then be less likely to be disfavored in hiring if both parents had to share the burden more equally. This is also crucial to retaining more young women in science. These rules should also apply to LGBT+ couples.
* *Add spousal considerations to hiring processes.* Institutions should recognize the existence of the dual-career situation and choose to deal with it, since half of the women with a PhD in physics have a spouse with a similar educational level (as opposed to only 20% for men). Institutions should take action before beginning a search, to provide assistance for spouses. This would help young women find positions without putting their relationships under strain. The same should also apply to LGBT+ couples.

To retain more members of minorities in science, we could:

* *Provide mentors for young people starting their careers.* The mentor should be different from their boss or supervisor to avoid any conflict of interest, and have proper institutional support. The mentor could, for example, make sure that young people progress properly, that they are given adequate funding and support, and that they get to attend meetings and give talks at various conferences. The mentor should be able to advise young people on academic and professional issues. These mentors should support everybody and, in particular, members of minority groups.
* *Have broad discussions about gender issues at large scientific meetings.* For example, men are often unaware of the situation faced by women in science and

continued

How to attract, hire and retain minorities in science (*continued*)
....

lack opportunities to discuss this situation, even though they are very often open to doing so. Members of majority groups often unconsciously discriminate against minorities. Education would help. Every majority group would benefit from learning more about the difficulties and specificities of other groups.

• *Hold scientific meetings for members of minority groups* where young people could see how valuable the contributions of members of their groups are, and where they could find positive reinforcement, get to talk with peers and get support. This would also provide a place for discussions on issues facing young women, as well as opportunities to share experiences and support each other. Women's groups, LGBT+ and black physicists' associations, etc. should be facilitated and supported.

Are women treated on an equal footing?

The representation of women, that is, the percentage of women, is not the only indicator that can be used to estimate whether female physicists are treated equally. The American Institute of Physics conducted a vast survey[8] in 2012 involving 15,000 physicists from 130 different countries. The goal was to compare the working experiences of men and women. This huge statistical sample had the immense advantage of providing a picture of the situation unbiased by personal perceptions. In Table 9.2, I summarize a large portion of this research. The results speak for themselves.

The first column gives the categories of questions asked. The participants were divided into two groups, namely scientists from less developed and from very developed countries. For each group, the table shows the percentages of women and men who answered "yes" to several questions. I have presented the averages of these answers after regrouping them into broad categories.

The first category of questions asked if the person had opportunities to take part in various activities such as attending conferences, being a speaker or invited speaker at a conference, conducting research abroad, serving on editorial committees for peer-reviewed journals or on other

---

[8] http://www.aip.org/statistics/reports/global-survey-physicists

Table 9.2 Compilation of the results of a vast survey conducted by the American Institute of Physics, involving 15,000 physicists from 130 countries.

Percentage of "yes"	Less developed countries		Very developed countries	
	Women	Men	Women	Men
Access to professional activities	50%	62%	50%	58%
Sufficient resources	40%	51%	48%	58%
Career affected by children	58%	50%	53%	41%
Assumed domestic tasks	39%	17%	44%	24%
Fewer challenges for parents	27%	9%	21%	4%

Source: Pauline Gagnon, based on the data from the American Institute of Physics.

important committees, and supervising students or their thesis work. In other words, the participants were asked if they were participating in activities that allow researchers to progress in their career. On average, 50% of all female physicists answered "yes" to these questions, compared with approximately 60% of male physicists, regardless of whether the participants came from less or very developed countries.

Women also reported being disadvantaged compared with their male colleagues in terms of having sufficient resources. This included having enough office and laboratory space, equipment, research budgets, travel budgets to attend conferences, and technical and administrative support. More women than men answered that their career had been affected after the birth of a child. In fact, this study showed that men with children were favored, while women with children were penalized. This supports the idea that men are still often perceived as the "breadwinner." The women also took on more domestic tasks and were offered fewer challenges at the professional level after they became mothers. In all these aspects, the table shows that there is a statistically significant difference in the treatment of the two sexes, that is, women are not treated on an equal footing.

This situation, although sobering, is evolving positively. More and more women are assuming high-responsibility positions in particle physics experiments. For example, Persis Drell was Director General of SLAC, a laboratory located in California, from 2007 to 2012. Fabiola Gianotti (Figure 9.4), spokesperson of the ATLAS Collaboration from 2009 to 2013, and heading three thousand people, became

Figure 9.4 Fabiola Gianotti, a former spokesperson of the ATLAS Collaboration, presenting the results on the discovery of the Higgs boson on July 4, 2012. Dr. Gianotti became CERN Director General in January 2016.
Source: CERN.

CERN Director General in January 2016. She is the first woman and the youngest person ever to fill this position. Also, Young-Kee Kim was Deputy Director of Fermilab, a laboratory near Chicago, from 2006 to 2013. But even more striking is the essential role played by more and more women in all particle physics collaborations, taking part in the day-to-day running of these experiments and contributing at all levels.

The LGBT+ community

Statistical studies and other types of investigations can help to identify problems that exist and find solutions. Unfortunately, there is no information about other underrepresented groups in physics, either in general or at CERN. These include people with disabilities, people of different races or religious beliefs, and members of the LGBT+ community. For the latter group, there has been a dynamic and active association of LGBT+ people at CERN since 2010 (Figures 9.5 and 9.6), although it initially encountered some resistance to gaining recognition.

Figure 9.5 Some members of the LGBT+ group at CERN gathered to produce a video for the series *It Gets Better*, addressing young people from sexual minorities. I am in the center of the first row.

Source: Yury Gavrikov.

Figure 9.6 Some members and friends of the CERN LGBT+ group visiting the ATLAS cavern.

Source: Yury Gavrikov.

Nevertheless, the community of physicists is generally open-minded, even if sometimes not too enlightened on the subject: for example, the posters of the LGBT+ group at CERN are regularly removed or vandalized. Such narrow-minded attitudes from individuals contribute to isolating the group instead of allowing it to bloom. Several CERN employees do not dare reveal their sexual orientation for fear of possible isolation or a change in the way they would be perceived in their work environment. Nevertheless, about 60 people participate in the activities of the LGBT+ group.

Scientists from the various experiments share their results at international conferences. These meetings are essential to keeping up to date with recent developments, and for networking and becoming known. It is thus particularly important that these meetings take place in countries where the physical safety of LGBT+ people is not threatened.

LGBT+ scientists face specific difficulties related to their work. For example, institutes in various countries send members of their staff to work at CERN, often for several years at a time. We need to make sure that the partners of LGBT+ staff can obtain visas for the duration of their stay. Otherwise, the sacrifices required are enormous, from the emotional, personal and financial viewpoints. Such situations impact negatively on the performance of the individuals affected. Most institutes and CERN itself also pay an allowance for couples and families. These allowances must also be offered to all couples and not be reserved for married couples, since only a few countries grant this privilege to same-sex partners. Even though both of the host countries of CERN, namely France and Switzerland, recognize unions between same-sex partners, the situation is still complex for foreigners. But the biggest problem remains homophobia. It is difficult to be well integrated into their workgroup if LGBT+ people feel that their coworkers will be ill at ease just because they have mentioned their same-sex partner, or if they are afraid that there could be discrimination in hiring. This last point is particularly important, since most researchers at CERN are on short-term contracts. Fortunately, mentalities are evolving, as they are everywhere else. One of the most effective weapons against homophobia is to refuse to stay "in the closet" and, instead, to live openly as an LGBT+ person. People with well-established professional reputations can greatly support younger people. If nobody hides, it means there is nothing to be hidden. Since homophobia often originates from a fear of the unknown, identifying oneself publicly as an LGBT+ person can help to dissipate such fears.

Racial diversity at CERN

The diagrams in Figures 9.7 and 9.8 give an overview of racial diversity at CERN. I have regrouped all CERN users, according to their nationality, into five big categories corresponding to the five continents. As of September 1, 2014, 72% of all users came from Europe. There were also more Asian scientists (13%) than North Americans (11%). South Americans accounted for 2%, whereas Africans accounted for a very thin 0.7% (Figure 9.9). The percentages in white type in each diagram show the proportion of women in each region.

Of course, CERN was initially established as a European laboratory and its member states are still essentially European countries to this day (except for Israel), hence the strong domination of European scientists. However, in recent years, CERN has started to take on the role of international leadership in particle physics. The two diagrams illustrate that

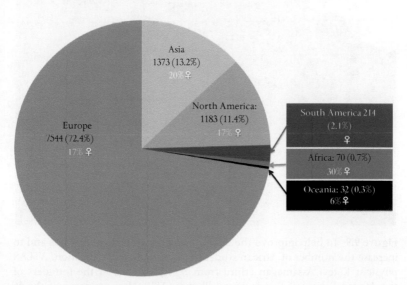

Figure 9.7 Repartition of CERN users by nationality, and percentage of women in each area.

Source: Pauline Gagnon/CERN.

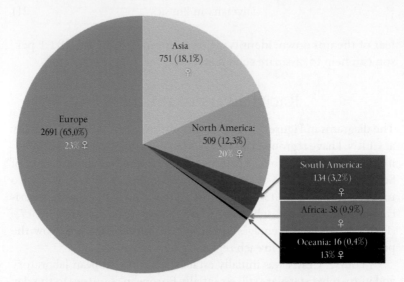

Figure 9.8 The distribution for CERN users under 35 years of age corresponding to Figure 9.7.

Source: Pauline Gagnon/CERN.

Figure 9.9 To help improve the quality of higher education in Africa and to increase the number of African students acquiring higher education, ATLAS physicist Ketevi Assamagan (third from right) was one of the founders of the African School of Fundamental Physics (ASP). Here are some of the 56 participants to the third school held in Dakar, Senegal in 2014.

Source: ASP 2014

the effects of the opening up of CERN to other countries and the efforts that have been made to train more people from other continents are starting to show.

Through its summer student program, CERN also offers opportunities to young people of all nationalities to take part in its research program while also attending a series of specialized lectures. However, the number of places offered to people from nonmember state countries is extremely limited. Furthermore, there is an absence of role models for these students. For example, between 2010 and 2013, only 14% of the lecturers were women, while 33% of the participants were women. The other minorities were also greatly underrepresented among the lecturers.

Figure 9.8 illustrates what the trend will be in coming years. This is based on the numbers of CERN users who are under 35 years of age today. The diagram illustrates how the composition of the scientists will change, with fewer Europeans and more Asians in the coming years. The proportion of Africans will also increase slightly, since young people make up half this group.

Women and the Nobel Prize in Physics

History has not been kind to women in science, and the awarding of Nobel Prizes is no exception. Marie Curie and Maria Goeppert-Mayer are, to this day, the only women to have received a Nobel Prize in Physics. Madame Curie is also the only person to have received two Nobel Prizes in different categories, namely physics and chemistry. There are unfortunately several notorious cases, and some others that are more debatable, of women who have been unfairly passed over. Here are some examples.

The story of Lise Meitner is probably one of the most striking cases of injustice. She was born in Austria in 1878 to a Jewish family. She received a doctorate in physics from the University of Vienna in 1906. Since, at the time, women were not allowed to hold academic positions in Austria, she left for Berlin in 1907 and began a collaboration with chemist Otto Hahn that lasted for more than 30 years. She was appointed head of a physics laboratory in 1917, a post she occupied until July 1938. She was then forced to go into exile in Sweden in order to escape from the Nazi regime, since she was Jewish.

---

continued

Women and the Nobel Prize in Physics (*continued*)

Nevertheless, she was able to meet secretly with Otto Hahn in November 1938 to discuss how to pursue their experiments, which Hahn conducted successfully with another collaborator. The two published their results shortly afterwards. The Nazi regime forbade Jewish people to put their names to scientific papers, and so Meitner's name could not appear on this article. Consequently, the Nobel Committee decided to award the Nobel Prize in Chemistry to Otto Hahn alone in 1944 for the discovery of nuclear fission. Shortly afterwards, several scientists realized the military potential of this discovery. Meitner was then invited to join the Manhattan Project in Los Alamos, the project that led to the development of the atom bomb. She refused, saying that she did not want to have anything to do with bombs. The scientific community today, however, now recognizes her contributions. The highest distinction of the Division of Nuclear Physics of the European Physical Society is the Lise Meitner Prize.

More recently, the Nobel Committee also passed over Jocelyn Bell-Burnell. Born in Northern Ireland in 1943, she obtained a doctorate in astronomy from Cambridge University in 1969. She participated in the construction of a radio telescope to study quasars, very energetic objects that emit radio waves and visible light. She noticed some faint, pulsed radio signal in her data and decided to investigate the origin of this pulsation in more detail, despite the lack of interest shown by her supervisor, Antony Hewish. He thought she was wasting her time, being convinced the signal came from some kind of interference or human source.

Her perseverance allowed her to discover the existence of pulsars, neutron stars that emit a pulsed signal. Nevertheless, Hewish and Martin Ryle, another member of their team, received the Nobel Prize in Physics for this discovery in 1974, provoking outrage in the astronomy community. The fact that she was a student at the time may have played a role. If this was the case, the Nobel Committee's attitude must have evolved since then: in 2010, Konstantin Novoselov and his supervisor, André Geim, jointly received the Nobel Prize in Physics for their discovery of graphene.

The case of Mileva Marić Einstein, the first wife of Albert Einstein, is much more debated, however, since the evidence is circumstantial. The publication of the first biographies of Mileva and Albert in the 1960s raised some doubts about the matter. These books were based on numerous testimonies

Women and the Nobel Prize in Physics (*continued*)

collected from their close friends and family, since few written documents in the hand of either Albert Einstein or Mileva Marić were available at the time.

But, subsequently, their son Hans Albert discovered a box that had belonged to her mother and contained part of the correspondence between his parents. These letters were made public in 1987, revealing numerous elements suggesting the existence of a scientific collaboration between the two of them. More recently, in 2006, the archives containing Albert Einstein's personal documents for the period from 1921 to 1955, kept at the Hebrew University of Jerusalem, were at long last opened to researchers. The evidence of all these documents converges to show that the couple worked jointly on several subjects, including the theory of relativity and the photoelectric effect. The latter brought Albert Einstein the Nobel Prize in Physics in 1921. Their divorce agreement, written in 1919, stipulated that Albert was to surrender the complete amount of the Nobel Prize to Mileva if that prize was ever awarded to him. He got the glory; she got the money.

Given the controversy surrounding Mileva Marić Einstein's scientific contributions to her husband's work, I refer you to Appendix B at the end of this book. I have collected the main facts and some opinions together there to allow you to form your own opinion.

Mentalities and society have since evolved. Let's just hope that these kinds of injustices will not happen again.

THE MAIN TAKE-HOME MESSAGE

We are far from real diversity or real equality between men and women in particle physics, although the situation is evolving positively. If the current trend is maintained, a better balance at every level should be reached by the year 2798 . . . There are, nevertheless, many changes that could easily be made to help improve the situation, to attract and retain a broader diversity of people in science. Since diversity means more creativity, science has everything to gain from including more people from currently underrepresented groups. Thanks to its position as a scientific leader, CERN has both the ability and a moral obligation to set an example at every level.

10

What Could the Next Big Discoveries Be?

I am neither a fortune-teller nor particularly gifted in long-term predictions but, like the majority of particle physicists, I expect fast developments, even revolutionary ones, in the next 10 or 20 years. As history shows, every time accelerators have increased their energy reach, it has always resulted in spectacular progress.

As things stand, the first period of data taking at the Large Hadron Collider, Run I, was extremely successful, with the discovery of the Higgs boson, despite the fact that the accelerator operated at an energy (8 TeV) lower than initially planned. With the restart of operations in 2015 at higher energy (13 TeV) and with higher intensity, there is everything to hope for. What, then, are the discoveries most expected in the coming years?

Already, at the end-of-the-year seminar in December 2015, the CMS and ATLAS Collaborations both reported having found a few events that could possibly reveal the presence of a new boson with a mass around 750 GeV, that is, six times the mass of the Higgs boson. Due to the difficulties inherent to the restart of the LHC at higher energy, the amount of data collected at 13 TeV in 2015 was five to seven times smaller than what had been accumulated in 2012 at 8 TeV by ATLAS and CMS. Hence, the experimentalists exerted much caution when they presented these results: small data samples are always prone to statistical fluctuations. But theorists, who have been craving for signs of new physics for decades, jumped on it. By mid-August 2016, 540 theoretical papers had been published to suggest just as many possible different interpretations for this still undiscovered new particle. The effect disappeared in the summer 2016 when more data became available, revealing it had just been a statistical fluctuation. All this excitement clearly illustrates how much physicists are hoping for a huge discovery in the coming years. It could easily have been like with the Higgs boson, which was officially discovered in July 2012 but had already given some faint signs of its presence a year earlier. At the beginning of 2016, there was just not enough data to tell. And with insufficient data, it is as if we are trying to guess if the train is coming by looking in the far distance on a foggy day. Only time can tell if the indistinct shape barely visible above the horizon is the long awaited train or just an illusion. Updates on new developments will be posted on my website.[1]

The LHC strategy for the next 20 years

During the first operation period, (denoted Run I), both the CMS and the ATLAS experiments collected 25 inverse femtobarns (fb^{-1}) of data at 7 and 8 TeV, where the inverse femtobarn is the unit used to measure the quantity of data. This amounts to approximately 2.5 million billion events. The LHC restarted in the spring of 2015 at higher energy for a second period of data taking (Run II), after the first prolonged technical stop, Long Shutdown 1 (Figure 10.1). By December 2015, ATLAS and CMS had collected only a few

Figure 10.1 Part of the technical team who participated in the consolidation of the Large Hadron Collider during the first long technical shutdown in 2013–14. This picture was taken after the completion of the 1695th section of the accelerator. This major program of work enabled the LHC to reach an energy of 13 TeV in 2015, that is, nearly twice the operating energy of 8 TeV reached in 2012.

Source: CERN.

---

[1] Follow me on twitter @GagnonPauline or see my website for updates on what is happening in the field : http://paulinegagnon3.wix.com/boson-in-winter#!blogs/c112v

The LHC strategy for the next 20 years (*continued*)
....

inverse femtobarn of data at 13 TeV. Run II will last until the end of 2018 and should bring four times more data than for Run I, namely 100 fb^{-1}. The strategy adopted thus consists of alternating running and maintenance periods over the next 20 years. The data sample is expected to triple between 2021 and 2023 during Run III getting to 300 fb^{-1}, to finally reach 3000 fb^{-1} by the end of Run IV around 2037. There will be enough data to suit everyone's taste.

Why not operate the accelerator continuously and maximize the number of events collected? The idea instead is to operate the accelerator at top capacity for approximately three continuous years, and then stop for about two years to increase the power of the machine and perform all of the inevitable maintenance work. This way, the experiments can also take advantage of the interruption to replace or repair any damaged subdetectors, and install improved ones when needed.

Each break also gives a chance for the experimentalists to complete their analysis of all the data collected during the previous data-taking period and to prepare for the next stage. For example, before each new run, it is

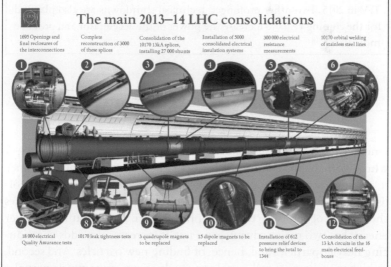

The main 2013–14 LHC consolidations

1 — 1695 Openings and final reclosures of the interconnections

2 — Complete reconstruction of 3000 of these splices

3 — Consolidation of the 10170 13kA splices, installing 27 000 shunts

4 — Installation of 5000 consolidated electrical insulation systems

5 — 300 000 electrical resistance measurements

6 — 10170 orbital welding of stainless steel lines

7 — 18 000 electrical Quality Assurance tests

8 — 10170 leak tightness tests

9 — 3 quadrupole magnets to be replaced

10 — 15 dipole magnets to be replaced

11 — Installation of 612 pressure relief devices to bring the total to 1344

12 — Consolidation of the 13 kA circuits in the 16 main electrical feed-boxes

Figure 10.2 Details of the LHC consolidation work undertaken in 2013–2014.
Source: CERN.

continued

The LHC strategy for the next 20 years (*continued*)

necessary to generate phenomenal quantities of simulated events corresponding to the new operating conditions. These events are essential for determining the selection criteria for the various analyses.

The first technical shutdown, in 2013–14, was needed not only to perform the extensive maintenance needed after three years of operation, but also to undertake a vast consolidation program (Figure 10.2). This enabled the LHC to reach its nominal energy and luminosity, that is, the intensity initially planned for the beams. The luminosity measures the number of protons in the beams per square centimeter and per second. The denser the beams, the higher the odds of producing collisions.

From 2010 until 2012, the LHC operated at approximately 75% of its nominal luminosity and at a lower energy, namely 8 TeV instead of the 14 TeV anticipated. This reduction in power was necessary to avoid another incident like the one that caused considerable damage to the accelerator and prevented operation for more than a year, 10 days after the LHC started up in 2008. The first prolonged technical shutdown was thus used mainly to improve the interconnections between the superconducting magnets (the cause of the incident in 2008) and to allow running at an energy of 13 TeV in 2015. Two other prolonged technical shutdowns are also planned for the coming years to increase the power of the accelerator and produce more data.

Prediction 1: Discovering or ruling out supersymmetry

One of the most anticipated discoveries is probably supersymmetry. A whole world of possibilities opened up with the restart of the LHC, since it can now produce more collisions, and at higher energy. More collisions mean more data collected by the experiments, which increases the chances of observing the rarest phenomena, and operating at higher energy has two major advantages. First, it increases the odds of producing heavier particles and, hence, of finding new particles. And second, at higher energy, we will be able to produce supersymmetric particles in larger quantities, assuming they exist, of course. We thus win on two fronts, increasing both the quantity of data and its reach.

If the theory of supersymmetry is really the theory that corresponds to "new physics," that is, all phenomena going beyond what is described by the Standard Model, we shall have a better idea about it toward the end of the LHC program during the fourth and last data-taking period, Run IV (see box "The LHC strategy for the next 20 years"). This last phase, called High Luminosity LHC, should take place around 2027–2037. As we saw in Chapter 6, the ATLAS and CMS experiments have already eliminated many possibilities for the masses of the various supersymmetric particles by exploring dozens of different scenarios. If these particles exist and if they are not too heavy (i.e., they are within the reach of the LHC), we could soon have the immense pleasure of discovering the first supersymmetric particles.

And if no discoveries are made, we shall at least have the satisfaction of having looked practically everywhere we could. Figures 10.3 and 10.4 show two diagrams illustrating what could be excluded in the next

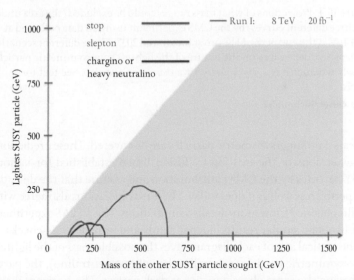

Figure 10.3 Mass values already excluded (the area under the three different curves) for the lightest supersymmetric particle after analysis of all data collected up to 2012 at 8 TeV by the CMS experiment. Three different scenarios are depicted, depending on the nature of the lightest supersymmetric particle, namely whether it is a *stop*, a *slepton*, or a *chargino* or *neutralino*. See text for more details.

Source: Oliver Buchmüller.

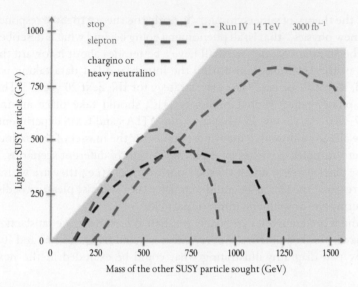

Figure 10.4 Projections of what mass ranges could be excluded (the area under the three different curves) by the CMS experiment using all data collected at 13 or 14 TeV at the end of the LHC program around 2037. Three different scenarios are depicted, depending on the nature of the lightest supersymmetric particle, namely whether it is a *stop*, a *slepton*, or a *chargino* or *neutralino*. See text for more details.

Source: Oliver Buchmüller.

20 years if no supersymmetric particles are discovered. These predictions are extensions of the existing exclusion limits established for various SUSY particles by the CMS Collaboration and assume that the detector will perform as well as it does today. These extrapolations also agree with results obtained from more detailed simulations. The ATLAS experiment should obtain similar results independently and provide cross-checks.

The vertical axis of each diagram gives the possible mass of the lightest supersymmetric particle (generally, the lightest neutralino), the particle having the same characteristics as dark matter. The horizontal axis gives the mass of another, heavier supersymmetric particle that could be produced by the LHC and would decay into the lightest SUSY particle.

Three scenarios are considered here: the cases where the lightest particle would come from the decay of a *stop* (red curve), a *slepton* (blue curve) or a *chargino* (black curve). The zone underneath each curve shows the values that have already been excluded for these three particles using all currently available data. Only the values in the shaded zone, below the diagonal, are allowed since, by definition, the lightest SUSY particle must be lighter than the other particle. These diagrams were produced assuming the particles studied (*stop, slepton, chargino* or *neutralino*) always decayed to produce the lightest SUSY particle. Figure 10.3 shows what has already been achieved during the first data-taking period, Run I, which ended in December 2012. By the end of 2015, there were still not quite enough data accumulated at higher energy to significantly modify this picture. The red curve delimits the set of values eliminated by the current searches for the *stop*. For example, the possibility of finding a *stop* with a mass up to roughly 600 GeV (five times the mass of the Higgs boson) if the mass of the lightest supersymmetric particle is below 250 GeV is already ruled out. The blue and black curves show the excluded regions for sleptons and charginos.

The dotted lines in Figure 10.4 illustrate which zones will be excluded by about 2037, at the end of the last data-taking period, Run IV, when we expect to have accumulated roughly 150 times more data, at nearly twice the energy of the first data-taking period. If SUSY has still not been discovered by then, much broader regions will have been excluded for the possible masses of the three types of supersymmetric particles shown in this plot. Theorists will then have much more information to constrain their theories, helping them steer their searches in the right direction.

Prediction 2: More information on the exact nature of the Higgs boson

The explosion in the sample size of the data from the CMS and ATLAS experiments in the coming years will bring increased precision to all measurements. It will be much easier to study all the properties of the Higgs boson and check in great detail whether everything agrees perfectly with the theoretical predictions. But these high-precision studies might also reveal small deviations from the predictions of the Standard Model. For example, the current measurements of the various production and decay channels of the Higgs boson carry experimental uncertainties ranging from 25 to 30%. These error margins will decrease to approximately 5% at the end of data taking around 2037. We shall then know much more about this boson.

Table 10.1 Number of events produced by the LHC and collected by each experiment at present and in the future.

	Run I	Run II	Run III	Run IV
Data-taking period	2010–13	2015–18	2021–23	2027–37
Data collected (fb^{-1})	25	100	300	3000
Collision energy (TeV)	7–8	13–14	14	14
Number of Higgs bosons produced	660,000	6 million	17 million	170 million
Number of Higgs bosons collected	25000	15000	4500	450000

When the LHC started operating at higher energy, the probability of producing heavier Higgs bosons, such as those predicted by supersymmetry, increased. By 2019, we shall already have enough data to determine whether there are more types of Higgs bosons or not. It will then be possible to exclude, within some SUSY models, the existence of Higgs bosons with a mass of up to 1000 GeV (or 1 TeV), that is, eight times the mass of the Higgs boson already found. This will also be sufficient to constrain or eliminate models that predict masses less than 1 TeV for the Higgs bosons associated with supersymmetry. In other words, the whole picture will become clearer.

Table 10.1 presents a summary of the operational characteristics of the LHC for the coming years. It gives an overview of what is expected to be achieved during the four data-taking periods planned up to about 2037. The inverse femtobarn (fb^{-1}) is the unit used to measure the volume of data collected by each experiment.

All these events will be extremely useful, since there are still so many unanswered questions regarding the Higgs boson discovered in 2012: Is it a composite particle or a fundamental particle? Is it the unique Higgs boson or one of many? Is it the first supersymmetric particle to be discovered? Does it establish a link between Standard Model particles and dark matter particles? Is it the cause of the difference between matter and antimatter? Did it cause the initial inflation phase of the Universe immediately after the Big Bang? All these questions require huge samples of Higgs bosons. We may then have a chance to answer some of these questions.

Prediction 3: The first anomaly in the Standard Model

All particles predicted by the Standard Model have now been found. Finding a new boson or any new particle would be the simplest way to discover what theory lies beyond the Standard Model. Another indirect way would be to measure a small deviation in a quantity predicted by the Standard Model. This is why three LHC experiments, CMS, ATLAS and LHCb (which specializes in this topic), are interested in very high-precision measurements involving b quarks. Physicists are particularly interested in studying the differences between matter and antimatter. This can be accomplished by studying how b quarks and their counterparts, b antiquarks, decay. Physicists are trying to understand why antimatter practically disappeared from the Universe when all of the results obtained in the laboratory indicate that matter and antimatter should have been produced in equal amounts after the Big Bang. One favorite method of looking for very small differences between the two is to study the rarest of the decays involving b and anti-b quarks. The slightest deviation would then be detectable. To date, some predictions of the Standard Model have proved to be exact up to the ninth decimal place for some particular measurements. This means that the experiments had to examine billions of events to obtain such precision.

The colossal quantity of data that will become available in the coming years will allow tests of unprecedented precision. Sooner or later, this should reveal a fault in the Standard Model. The discovery of any kind of anomaly would steer the theorists in the right direction and help them understand what the "new physics" is. There is nothing like a good experimental discovery to put the theory back on the right track.

Prediction 4: A little light on dark matter

The search for direct evidence of the existence of dark matter is undoubtedly one area where big developments are expected in the coming decade. This discovery alone would completely revolutionize particle physics and cosmology. After only three decades of activity, researchers have already managed to set limits a hundred thousand times smaller than those established 30 years ago for the probability of interaction between WIMPs, the hypothetical dark matter particles, and ordinary matter.

Large international collaborations are now being formed to search for dark matter. Such cooperation enables researchers to pool their material and scientific resources, and is yielding several new detectors. A new generation of more massive and more powerful detectors is currently under construction in Canada, the United States and Italy. These second-generation detectors will be sensitive to lighter WIMPs, thanks to a spectacular reduction in their background level. In particular, the SNOLAB laboratory is rapidly growing 1 mile (2 km) underground in the Vale Creighton mine near Sudbury in Canada. There, an international team is assembling the SuperCDMS detector, which should start operation some time soon. This detector will be able to detect ultralight WIMPs, a still unexplored region (the top left corner of the graph in Figure 10.5).

This graph is a more elaborate version of the one presented in Chapter 5, which is already complex enough. It is hard to imagine a more

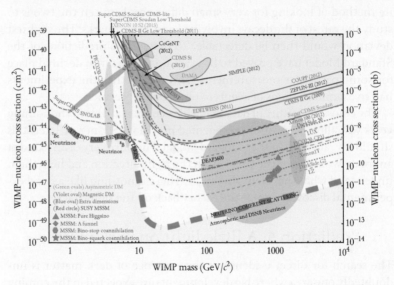

Figure 10.5 Projections of what could be known about dark matter particles in the coming decade. The whole area of the plot represents all possible values of the probability of a dark matter particle interacting with matter, depending on its mass given by the horizontal axis. The light green area above the solid green line is already excluded. All of the areas above the various dashed lines should be excluded by dark matter direct search experiments currently running or planned. If nature is so unkind that dark matter interacts extremely rarely (lower part of the plot) or if the dark matter particle is extremely light (left part of the plot), it will not be possible to discover dark matter with current experimental techniques. The broad orange dashed line delimits the region where the neutrino background drowns out the dark matter signal. The area below the light dashed lines will remain unexplored.

Source: Snowmass Community Summer Study 2013.

convoluted plot, but, nevertheless, it is relatively easy to extract the gist of it. The vertical axis shows the probability of an interaction between a dark matter particle and a particle of ordinary matter. This is measured in units of area (cm^2), since it represents the size of the target (a nucleon) as seen by an incoming dark matter particle. The larger the target, the more likely it is to hit it. The horizontal axis shows the possible mass values in GeV of dark matter particles.

There are three broad zones of interest. First, the part in green in the upper section of the plot corresponds to all values already eliminated by current experiments. Second, the zone in yellow at the bottom of the graph represents all values where the background from neutrinos dominates. It will not be possible to look for dark matter particles in this zone with current experimental techniques. Third, the area in white contains all values where the current techniques are not blocked by the neutrino background, but which the experiments currently ongoing have not yet been sensitive enough to explore. In this version of the graph, new curves (the various dashed lines) have been added in this zone. Each of these represents an experimental limit that the various experiments hope to obtain in the coming years. All values above these curves will then be excluded. Hence, in the coming decade, a substantial fraction of all places in which we can look with the current experimental techniques will have been explored.

If none of the current or planned experiments has found dark matter particles by then, we will need new methods to pursue the quest for dark matter particles since the neutrino background will start interfering. New avenues are already under study to push back the limits imposed by the neutrino background. For example, one could eliminate all neutrinos coming from the Sun by taking into account the direction of the particles striking the detector.

Good news will not necessarily come from underground. Evidence, albeit indirect, could also come within a few years from the AMS-02 experiment on board the International Space Station. The team will most likely by then have completed accumulating and analyzing enough data to shed some light on the origin of the positrons found in cosmic rays. Do these positrons come from conventional astronomical sources such as pulsars, or could they be the first signs of interactions between dark matter particles and ordinary matter, as discussed in Chapter 5? We should know more within a few years, although some theorists doubt whether the AMS-02 data will be precise enough to be decisive. In the longer term, the LHC, with 150 times more data, will continue to look for dark matter particles in the decays of Higgs bosons, in the form of the lightest supersymmetric particle as well as many other possibilities.

The medium and long-term future

Over the past few years, a radical change has taken place in the approach adopted by the countries involved in particle physics. Everyone now realizes that no country on its own can afford the sophisticated tools, namely detectors and accelerators, needed today. International co-operation has therefore become the norm, in order to pool the human, technical and financial resources needed for such megaprojects. CERN is playing a more central role in the international community and is continuing to invite new countries to join its ranks.

Particle physics has become a symbol of international cooperation. Several new accelerator projects are already under study (Figure 10.6) to take over from the Large Hadron Collider when it retires around 2037.

Figure 10.6 The linear compact collider (CLIC) is a project at CERN currently under study as a possible follow-up project to take over from the LHC. CLIC could generate high-energy electron–positron collisions, thanks to a primary beam of low energy but high intensity.

Source: CERN.

No option has yet been chosen definitively, but all countries have agreed to work within international collaborations.

So, what is the most extraordinary discovery one can hope for from all these detectors and accelerators? A new particle that would reveal the nature of the "new physics" beyond the Standard Model? The confirmation of a theoretical hypothesis such as supersymmetry, or the discovery of dark matter particles? An unexpected, complete surprise? All would be great. No matter what might be revealed, be it predicted or not, the chances of new discoveries are enormous at the moment, as has proved to be the case in the past every time accelerators increased in energy. This imminent prospect is terribly exciting and the atmosphere is feverish, since we are about to break new ground. This is motivating thousands of physicists involved in particle physics today. Humanity will soon be able to go to bed knowing a little more.

THE MAIN TAKE-HOME MESSAGE

The Large Hadron Collider is still in its early stages. Over the coming 20 years, it will produce enough data to allow enormous progress in particle physics. New particles or anomalies could soon be discovered, revealing at long last the nature of the "new physics" beyond the Standard Model. If supersymmetric particles exist, we will have every chance of discovering them. And if they prove not to be there, we shall have much more experimental data that we can use to formulate better theoretical hypotheses about new physics. We shall also finally start tackling some of the remaining questions about the Higgs boson: whether it is unique, whether it establishes a link between ordinary matter and dark matter, and whether it is supersymmetric or not. In the coming decade, the many ongoing and planned experiments will give us much more information about dark matter. We could even have the huge pleasure of catching the first dark matter particles. And, who knows, what we might discover could be completely unexpected. With the increase in energy and intensity of the LHC and the construction of second-generation dark matter detectors, the chances of new discoveries are tremendous.

Percentages of Women for the 101 Nationalities Present at CERN

CERN users by nationality	Percentage of women	Percentage of women below 35 years	Percentage of people below 35 years	Total number of people at CERN
Afghanistan	0%	0%	100%	1
Albania	50%	100%	50%	2
Algeria	20%	0%	20%	5
Argentina	31%	29%	44%	16
Armenia	23%	31%	59%	22
Australia	7%	13%	59%	27
Austria	11%	15%	33%	81
Azerbaijan	0%	0%	43%	7
Bangladesh	0%	0%	67%	3
Belarus	8%	8%	30%	40
Belgium	25%	25%	54%	109
Bolivia	33%	33%	100%	3
Bosnia	0%	0%	100%	1
Brazil	20%	12%	54%	111
Bulgaria	22%	44%	22%	74
Cameroon	0%	0%	100%	1
Canada	16%	22%	48%	141
Cape Verde	0%	0%	100%	1
Chile	21%	25%	86%	14
China	22%	23%	72%	302
Colombia	6%	4%	77%	35
Croatia	28%	32%	53%	36
Cuba	50%	57%	70%	10
Cyprus	17%	13%	89%	18
Czech Republic	9%	10%	51%	216
Denmark	9%	21%	36%	53
Ecuador	0%	0%	75%	4
Egypt	42%	73%	58%	19
El Salvador	0%	0%	0%	1
Estonia	20%	27%	73%	15
Finland	19%	21%	30%	79
France	17%	25%	26%	731

continued

CERN users by nationality	Percentage of women	Percentage of women below 35 years	Percentage of people below 35 years	Total number of people at CERN
FYROM	100%	100%	100%	1
Georgia	19%	13%	41%	37
Germany	14%	19%	47%	1095
Gibraltar	0%	0%	100%	1
Greece	28%	32%	38%	152
Hungary	12%	22%	34%	67
Iceland	0%	0%	25%	4
India	23%	26%	52%	214
Indonesia	0%	0%	29%	7
Iran	32%	41%	61%	28
Iraq	0%	0%	100%	1
Ireland	14%	13%	73%	22
Israel	15%	29%	33%	52
Italy	23%	31%	29%	1666
Japan	7%	8%	47%	253
Jordan	0%	0%	100%	1
Kazakhstan	100%	100%	100%	1
Kenya	50%	100%	50%	2
Latvia	0%	0%	100%	1
Lebanon	42%	42%	100%	12
Libya	100%	100%	100%	1
Lithuania	10%	8%	62%	21
Luxembourg	25%	50%	50%	4
Madagascar	0%	0%	33%	3
Malaysia	20%	20%	67%	15
Mauritius	0%	0%	100%	1
Mexico	19%	28%	58%	69
Montenegro	0%	0%	0%	3
Morocco	27%	25%	36%	11
Myanmar	0%	0%	0%	2
Nepal	33%	50%	67%	6
Netherlands	10%	28%	25%	144
New Zealand	0%	0%	0%	5
North Korea	0%	0%	0%	1
Norway	29%	33%	41%	59
Pakistan	12%	10%	49%	43
Palestine	40%	50%	80%	5
Peru	0%	0%	75%	8
Philippines	0%	0%	100%	1

continued

CERN users by nationality	Percentage of women	Percentage of women below 35 years	Percentage of people below 35 years	Total number of people at CERN
Poland	19%	16%	39%	247
Portugal	20%	21%	45%	104
Qatar	0%	0%	100%	1
Romania	26%	30%	36%	121
Russia	11%	18%	22%	951
Saudi Arabia	100%	0%	0%	2
Senegal	0%	0%	0%	1
Serbia	38%	47%	43%	40
Singapore	33%	33%	100%	3
Sint Maarten	50%	0%	0%	2
Slovakia	17%	21%	51%	102
Slovenia	20%	50%	30%	20
South Africa	28%	44%	50%	18
South Korea	19%	23%	49%	115
Spain	24%	31%	38%	323
Sri Lanka	25%	0%	50%	4
Sweden	24%	36%	39%	71
Switzerland	14%	18%	31%	177
Syria	100%	100%	100%	1
Taiwan	20%	16%	54%	46
Thailand	33%	38%	67%	12
Tunisia	50%	50%	100%	4
Turkey	33%	40%	59%	159
Ukraine	10%	14%	58%	60
United Kingdom	12%	17%	46%	633
United States	14%	18%	41%	973
Uzbekistan	20%	0%	20%	5
Venezuela	40%	44%	90%	10
Vietnam	36%	40%	91%	11
Zimbabwe	33%	33%	100%	3

Source: Pauline Gagnon, based on CERN data as of September 1, 2014.

The Role of Mileva Marić Einstein

In 1999, *Time* Magazine named Albert Einstein "Personality of the Century." Einstein's scientific output, especially during the year 1905, raises many questions. Scientists have wondered for decades how a single person could have published so many articles as a sole author. In parallel, but largely unknown to most scientists, biographers of his first wife, the mathematician and physicist Mileva Marić, have presented voluminous evidence attesting to her scientific contributions to her husband's work. What role did collaborative work play in Einstein's achievements? The question is particularly important given how essential collaborative efforts are in science today. In the case of Albert Einstein, the lack of irrefutable evidence and the absence of all of the people directly involved make it difficult to come to a decisive conclusion. However, several testimonies and documents still exist today. They allow us to get a picture of Mileva Marić's contributions, although opinions still diverge on this subject. As we shall see, her tragic fate was not only determined by the reprehensible actions of her husband but also branded with the red-hot iron of her times. My goal is not to denigrate a famous man, but rather to examine the possible contributions of his wife by analyzing existing documents and considering them in the social context of their time.

Until quite recently, historians had access to only part of Einstein's personal documents, those covering the period from 1879 to 1921. But in 2006, the archives of the Hebrew University of Jerusalem, containing Einstein's personal documents from 1921 to 1955, were at long last opened to researchers. This enabled Radmila Milentijević, a professor of history at the City University of New York, to paint a more complete portrait[1] of Mileva Marić and to fill in numerous gray areas of her role, in both the emotional and the scientific life of Albert Einstein.

A few books had previously been devoted to Mileva Marić, such as, for example, the first biography,[2] written by Desanka Trbuhović-Gjurić and published in 1966. It was reedited in 1999 to include Albert and Mileva's love letters, which were made public at the end of the 1980s. This book is based mostly on the

---

[1] Radmila Milentijević, *Mileva Marić Einstein: Life with Albert Einstein*, United World Press, 2015. All quotes have been taken from the French version (which appeared before the original version in English), *Mileva Marić Einstein—Vivre avec Albert Einstein*, Éditions de l'Age d'Homme, 2013.

[2] Desanka Trbuhović-Gjurić, *U senci Alberta Ajnstajna* (In the Shadow of Albert Einstein), biography of Mileva Marić published in Serbian in 1969, translated into German in 1988 and reedited in 1999. The German edition was translated into French in 1991. All subsequent quotes refer to the French version, *Mileva Einstein, Une vie*, Editions des femmes.

testimonies of people who knew Mileva Marić. Milentijević had access to all the personal documents of Albert Einstein and Mileva Marić, yielding a more complete book. She also benefited from the wealth of researchers who had preceded her, such as Dord Krstić, a professor at Ljubljana University, who wrote an excellent book[3] based on numerous interviews he conducted with relatives and friends of Mileva over several decades. The pages that follow give a short summary of Milentijević's book, from which I have gathered various passages, facts and citations from other authors quoted in the book. I have also included information from several other sources.

Facts and testimonies

Radmila Milentijević's book traces Albert and Mileva's life together, from their meeting to the time of Mileva's death in 1948. Mileva Marić (Figure B.1), a native of Serbia, was 21 years old when she met Albert at Zurich Polytechnic School (ETH) in 1896, where they both studied physics. Albert Einstein was German and three years younger. Starting in 1899, a deep passion developed between them. They shared everything: their love, their studies, their research

Figure B.1 Mileva Marić when she was a student at the Polytechnic School in Zurich in 1897.

Source: Wikipedia.

---

[3] Dord Krstić, *Mileva & Albert Einstein: Their Love and Scientific Collaboration*, Didakta, Radovljica, Slovenia, 2004.

and their music. They started collaborating shortly after they met, as evidenced by numerous personal documents and numerous testimonies. The Albert Einstein Museum in Bern has preserved Albert's notebooks. Whole sections are written in Mileva's hand. Part of their correspondence from 1899 to 1903 still exists today. Mileva kept the 43 letters that Albert sent her, although the majority of hers were lost or destroyed. Only ten of them have been preserved.

From the beginning of their relationship, Albert regularly referred in his letters to the joy that working with Mileva brought him. She guided his reading and put some order into his life, helping him channel his bohemian temperament. Milentijević stresses how Einstein's letters are peppered with terms such as "our new studies," "our research," "our views," "our theory," "our article" and "our work on relative motion."[4]

Mileva and Albert studied together, and they obtained similar marks in their university courses. But in 1900, Mileva failed the final oral examination leading to the diploma. Was this due to a lack of competence or a sign of the prevailing attitude toward women at the time? Switzerland was then one of the few countries where women were admitted to university. Einstein passed this examination but unlike his three former schoolmates, did not find an academic position as he could have expected after his studies. During the following two-year period, having no resources, he returned periodically to live with his family while Mileva stayed in Zurich. The couple wondered in their letters if possible antipathy from one of Einstein's professors could explain why he was having such difficulty in obtaining his first position.

In a letter sent to her friend Helena Savić on December 20th, 1900, Mileva Marić mentions their collaboration, quoting a scientific paper on "the theory of liquids" that was published in March 1901: "We also sent a copy to Boltzmann and we would like to know what he thinks of it. I hope he'll write to us."[5] Although this comment suggests that she had participated in the work on the paper, it was published under the name of Albert Einstein alone.

Milentijević comments:

> One can conclude from this, that in spite of the fact that this paper was the product of a collaboration, Mileva and Albert decided that it would be published under Albert's name only. Why? Albert was unemployed. His personality and behavior at Polytechnic School seriously impeded his chances of getting a position. The only way to overcome this disadvantage for Albert was to demonstrate that he was a respected scientist and to establish his reputation within the academic community. For this he needed Mileva's help.[6]

[4] Milentijević, p. 13.

[5] Milan Popović, *In Albert's Shadow: The Life and Letters of Mileva Marić, Einstein's First Wife*, The Johns Hopkins University Press, Baltimore & London, 2003, letters from Mileva Marić to Helena Savić dated December 20, 1900, p. 70. Boltzmann was an eminent physicist of this period.

[6] Milentijević, p. 77.

Einstein himself gives evidence of their collaboration and the participation of Mileva in the theory of relativity in a letter dated March 27, 1901, in which he wrote to Mileva, "How happy and proud I will be when the two of us together have brought our work on relative motion to a victorious conclusion!"[7] This quote, written in Einstein's hand, constitutes the most direct proof that they worked together on the theory of relativity.

In May 1901, the fate of Mileva took a decisive turn. She became pregnant following a lovers' escapade with Albert in Como. Without a job, he did not want to marry her then, wanting instead to be able to provide for the needs of the household. Three months later, with the additional pressure of an uncertain future, Mileva again failed her oral examination. On December 28, 1901, Albert wrote to her, "When you will be my little wife, we shall resume diligently our scientific work, such as to not become Philistines."[8]

In the fall of 1901, Mileva went back to Serbia to stay at her parents' home. She came back briefly to Switzerland in October 1901 to try to convince Albert to marry her, but to no avail. At the end of January 1902, she gave birth to a girl named Liserl. Brokenhearted, she abandoned the child and came back twice to Switzerland to visit Albert. On July 1, 1902, Einstein finally obtained a subordinate position at the Patent Office in Bern, thanks to the intervention of his friend Marcel Grossmann's father. They got married in January 1903 but never took back their daughter. Did Einstein refuse out of fear that this illegitimate child could damage his career, a real threat at the time? This is at least an opinion suggested by Dennis Overbye, a scientific correspondent of the *New York Times* and author of a biography of Einstein. Overbye wrote, "As if in some Greek tragedy, then, the prize of their life would be their child. Mileva was too intelligent and introspective a woman not to be aware of the irony of fate, as she set about mortgaging her happiness to Albert's career."[9] No trace of their daughter was ever found. Milentijević thinks that she was probably given up for adoption in September 1903.

Dord Krstić wrote that Mileva's brother, Miloš Marić Jr., spent time during his medical studies in Paris and Bern. In 1905 "Miloš, who stayed with the Einstein family, had the opportunity to see 'close up,' exactly how Mileva and Albert lived and worked together."[10] Albert dedicated his time to their research after his office hours, and Mileva did so after completing her domestic tasks. She was also taking care of their first son, Hans Albert, who was born in 1904. Again according to Krstić, Miloš Marić reported back to his family and friends that the Einstein couple worked very hard. "He described how during the

---

[7] *The Collected Papers of Albert Einstein*, Princeton University Press, 1987–2006. Document 94, pp. 160–161.

[8] *The Collected Papers of Albert Einstein*, Document 131, pp. 189–190.

[9] Dennis Overbye, *Einstein in Love*, Penguin Books, New York, 2000, p. 91. (Quoted by Milentijević.)

[10] Krstić, p. 105.

evenings and at night, when silence fell upon the town, the young married couple would sit together at the table and at the light of a kerosene lantern, they would work together on physics problems. Miloš Jr. spoke of how they calculated, wrote, read and debated."[11] The author, Dord Krstić, comments that he heard this story firsthand from two relatives of the Marić family, first from Sidonija Gajin in May 1955, and then from Sofija Golubović in 1961.

To support the validity of Miloš Marić's testimony, Krstić takes great care over drawing a detailed portrait of Mileva's brother Miloš Marić, a renowned doctor. To establish his reliability, Krstić quotes colleagues of Miloš Marić describing him as a man of great integrity, scrupulous in sticking to the facts rather than inclined to follow his own opinions.[12]

Soon after Miloš Marić's visit in 1905, Albert Einstein published not only the theory of relativity but also four other scientific articles, including his doctoral thesis. One article described the photoelectric effect, work for which Einstein received the Nobel Prize in 1921. One can only wonder about this phenomenal productivity, especially from someone holding down a full-time technical job at the Patent Office. Hence, 1905 is referred to as Einstein's *annus mirabilis* ("miraculous year"). It was by far the most prolific period of his career.

Milentijević also quotes Peter Michelmore, another biographer of Einstein, who had several direct contacts with Einstein. Michelmore reports that after having spent five weeks of hard work on completing the article "On the electrodynamics of moving bodies," which laid the foundations for the special theory of relativity, "Einstein's body buckled and he went to bed for two weeks. Mileva checked the article again and again, then mailed it."[13] Afterwards, the couple went on holiday to Serbia to visit the Marić family. On this occasion, Mileva is reported to have told her father, "Before our departure, we finished an important scientific work which will make my husband known around the world."[14]

Krstić notes in his book that he heard this story from Mileva's cousin Sofija Galić Golubović in 1961. She was present when Mileva confided this to her father. Two other people, Simonija Gajin and Zarko Marić, repeated the exact same sentence separately in 1955 and 1961, respectively, both having heard it themselves from Mileva's father.[15]

Mileva's brother used to have many young intellectuals visiting him. Albert Einstein, attending one of these gatherings, declared: "I need my wife. She solves for me my mathematical problems." Mileva acknowledged this.[16] But

[11] Krstić, p. 105.

[12] Krstić, pp. 214–222.

[13] Peter Michelmore, *Einstein: Profile of the Man*, Dodd, Mead & Company, New York, 1962, p. 46. (Cited by Milentijević.) Also reported by Trbuhović- Gjurić, p. 103.

[14] Krstić, p. 115; Trbuhović-Gjurić, p. 105.

[15] Krstić, p. 115.

[16] Trbuhović-Gjurić, pp. 105–106.

around the end of 1912, the mathematician Marcel Grossmann seemed to have replaced her. Grossmann and Einstein jointly signed a first article on general relativity in 1913. Albert Einstein alone signed the second article on general relativity in 1915.

Trbuhović-Gjurić quotes Dr. Ljubomir-Bata Dumić, who wrote about his memories of the Einstein family's visit to Serbia in 1905:

> We were really looking up to Mileva, as if she was a divinity: her mathematical knowledge and her cleverness really impressed us. She could solve relatively simple mathematical problems on the spot, in her head, and what would have required several weeks for hard-working specialists she could solve in two days. And she always found the solution in her own way, original, the shortest way. We knew she had made him [Albert], that she was the author of his glory. She solved all his mathematical problems for him, especially those in relativity. As a dazzling mathematician, she dazzled us.[17]

Mileva's father, during his first visit to Switzerland, wanted to help the young couple financially and offered them a huge sum of money, 100,000 Swiss francs. Albert declined, saying:

> I did not marry your daughter for her money but because I love her, because I need her, because together, we are one. Everything I did and obtained, I owe it to Mileva. She is my brilliant inspiration, the angel who saves me from all sins in my life and even more so in science. Without her, I would not have started nor completed my work.[18]

Dord Krstić cites only the first sentence whereas Trbuhović-Gjurić quotes the whole paragraph but without specifying the source. Many authors, including Desanka Trbuhović-Gjurić, report that in 1908, Mileva collaborated with Paul Habicht, a student of Einstein, on the construction of an ultrasensitive voltmeter capable of measuring voltages as small as one ten-thousandth of a volt. This work took a long time because Mileva had numerous other commitments and constantly tried to improve the device. Trbuhović-Gjurić stresses that Mileva excelled at experimental work in the laboratory. "When they were both finally satisfied, they let Albert describe the apparatus, since he was an expert with patent applications."[19] She also relates that, when questioned by Paul Habicht's brother as to why her name did not appear on the patent, Mileva answered him with a pun, "Warum? Wir beide sind nur ein Stein" (literally, "Why? The two of us are only one stone" (ein Stein), i.e., we are only one).[20]

[17] Trbuhović-Gjurić, p. 106.
[18] Trbuhović-Gjurić, p. 107.
[19] Trbuhović-Gjurić, p. 95.
[20] Krstić, p. 115; Trbuhović-Gjurić, p. 95.

For the historian Radmila Milentijević, it is clear that Mileva, like many women of her time, chose to step back to allow her husband to succeed. He particularly needed her help after he failed to secure an academic post at the end of his studies like the three other students in his class. Clearly, for Mileva, they were one unique entity. Nevertheless, one year later, on September 3, 1909, Mileva expressed her first qualms to her friend Helena Savić: "[My husband] is now regarded as the best of the German-language physicists, and they give him lots of honors. I am very happy for his success, because he really does deserve it; I only hope and wish that fame does not have a harmful effect on his humanity."[21]

One often cited, but in my opinion, weak argument is the following anecdote. The physicist Abram Fedorovich Joffe was the assistant of Wilhelm Röntgen, a member of the editorial committee of the German journal *Annalen der Physik*, when the first article on the theory of relativity was published. In a eulogy for Einstein in 1955, he wrote[22] about having seen the original document of this first article on the theory of relativity, and that this article was signed with the joint name Einstein-Marity. *Marity* is the Hungarian version of Mileva's family name, as it appeared on her marriage certificate. Joffe explained at the time that Einstein had associated his name with his wife's name "according to the custom in Switzerland." However, there was no such custom, and only Mileva used this joint name.

For the physicist Evan Harris Walker,[23] Joffe could not have invented this story. If so, he would have transliterated the name Marić into Russian as МАРИЧ (*Maritch*, the phonetic equivalent of Marić) and not as МАРИТИ (*Marity*), which corresponded exactly to the Hungarian version of her official name in Switzerland. Walker concludes from this that Joffe had truly seen the article signed this way. The original article has since gone missing from the *Annalen der Physik* archives. In 1943, Einstein recopied the original article on the theory of relativity by hand for a charity auction, specifying on the copy that he had thrown the original away after publication.

Their divorce agreement, concluded in 1919, stated that, besides having to pay alimony for his children, Albert Einstein agreed to turn over the entire sum of money of the Nobel Prize to Mileva if that prize was ever awarded to him. All the money was eventually given to Mileva, although after much delaying and numerous reminders, as their correspondence revealed. In 1925, Einstein tried to establish in his will that this money constituted an inheritance for his two

---

[21] Popović, Mileva Marić's letter to Helena Savić, September 3, 1909, p. 98.

[22] A. F. Joffe, *Reminiscences of Albert Einstein*, published in *Uspekhi Fizicheskikh Nauk*, Vol. 57, No. 2, October 1955, p. 187. (Cited by Milentijević.)

[23] Evan Harris Walker, "Ms Einstein," unpublished article, reproduced by The Walker Cancer Institute Society, http://simson.net/ref/1995/MsEinstein.pdf.

sons, Hans Albert and Eduard. Mileva opposed this new arrangement, stating that this money belonged to her.

She seems to have intended then to provide evidence of her scientific contributions. But Einstein ridiculed her in a letter dated October 24, 1925, quoted by Milentijević:

> You really made me laugh when you began to threaten me with your memories. Had it ever occurred to you, if only for one second, that nobody would pay the slightest attention to your figments if the man about whom you speak had not achieved something important? When a person is completely insignificant, there is nothing else to tell such a person but to remain modest and silent. This is what I advise you to do.[24]

What happened to Mileva?

After the publication of all these articles in 1905, Albert Einstein's fame quickly expanded. He finally obtained several academic positions, first in Zurich and then in Prague. He and his family returned to Zurich (Figure B.2 depicts the couple around this time) and finally moved to Berlin in 1914, where he had begun an affair with his cousin, Elsa Einstein. Shattered, Mileva returned to live in Zurich with her two sons. Albert asked for a divorce, obtained it in 1919 and married his cousin.

Mileva survived thanks to mathematics and piano lessons that she gave to supplement the alimony she received from Albert, which he often delayed paying. Their son Hans Albert wrote to him on several occasions to remind him of

Figure B.2   Mileva Marić Einstein and Albert Einstein in 1912.
Source: Wikipedia.

[24] The Albert Einstein Archives, Hebrew University of Jerusalem. Letter from Einstein to Marić, October 24, 1925, AEA 75–364. (Cited by Milentijević, pp. 142–143.)

the difficult circumstances they were living in. With the Nobel Prize money, Mileva bought two rental properties and lived on the income from them. Their younger son, Eduard, born in 1910, suffered from schizophrenia and had to stay in a psychiatric hospital on many occasions from 1932 onwards. Einstein never saw him after 1933, the date on which he emigrated to the United States, but he kept in touch with Mileva all his life.

In spite of severe health problems and the difficulties generated by the war, Mileva dedicated all her energy and money to her sick son. In 1932, she asked Albert for a letter of recommendation so as to be able to obtain a teaching position in a girls' high school, so that she would be able to support herself and Eduard. Albert refused, saying that he "could make no recommendation for her at a moment when so many people younger than her were unemployed."[25]

When she got into debt owing to medical expenses for her son, her creditors threatened to take her house. Albert then agreed to buy back her house so that she and their son would not become homeless. Just before her death in 1947, she used a subterfuge to resell the house, even though Albert was the owner. She put all the money in Eduard's name to ensure continuous care for her son after her death. She died in 1948 in Zurich.

The opinions

John Stachel,[26] the first editor of *The Collected Papers of Albert Einstein*, where the letters exchanged by Albert and Mileva at the turn of the century were published, points out that the letters kept by Mileva Marić refer to her own research very little or not at all. However, the physicist Evan Harris Walker, in an article entitled "Ms Einstein," wrote "I find statements in thirteen of [Albert's] 43 letters to [Mileva] where reference is made to her research or to an ongoing collaborative effort."[27]

Quoting from Stachel's article in *Creative Couples in the Sciences*, Radmila Milentijević wrote "So Stachel attributed the use of the words 'we' and 'our work' [by Einstein in his letters to Marić] to the emotion physics provoked in [him], something he felt obliged to share with Marić." Stachel added, "The references made to common work were made at a difficult time of their relations, and were meant to reassure her on his love and his worship."[28] However, Milentijević,

[25] Letter from A. Einstein to M. Marić dated June 4, 1932, Albert Einstein Archives, Jerusalem, AEA 75–434, cited by Milentijević, p. 379.

[26] *The Collected Papers of Albert Einstein*, Princeton University Press, 1987–2006.

[27] Walker, "Ms Einstein," p. 7.

[28] John Stachel, "Albert Einstein and Mileva Marić: A collaboration that failed to develop," in *Creative Couples in the Sciences*, eds. Helena Mary Pycior, Nancy G. Slack and Pnina G. Abir-Am, Rutgers University Press, 1996, pp. 207, 209, 216. (Cited by Milentijević.)

who has traced the couple's life up to Mileva's death, refutes this argument, pointing out that "Albert was anything but an altruist."[29]

In another article, which appeared in February 1989 in *Physics Today*, the magazine of the American Physical Society, Evan Harris Walker concluded:

> Their years together saw Einstein's greatest achievements: His physics was filled with daring concepts of space and time distorted, of gravitation's being only a distortion of the space-time metric, of photons that truly were packets of energy—not just as a mathematical device as Max Planck thought, but as a reality. And his work was filled with the immediate implications of the most recent and detailed findings of the current physics. But after his marriage to Mileva ended, his physics became conservative. He added the cosmological constant to his equations so that they would predict the physics everyone expected for the universe, and as a consequence, he missed predicting the Big Bang. He became not the leader of the avant-garde physicists, but in time, the odd man out on his position against the new quantum theory.[30]

In 1929, a friend of Mileva, Milana Sefanovic, declared in an interview given to the Serbian newspaper *Politika*, that Mileva was "the most qualified person to speak about the genesis of the theory [of relativity] since she worked on it with him. Five or six years ago, she told me painfully about it. Maybe it was hard for her to be reminded of these happy times, maybe she did not want to cause any prejudice to her ex-husband."[31] In a letter sent to her friend Helena Savić, Mileva explained her position:

> Milana could not help confiding our stories to the newspaper reporter, and I thought then that the matter was finished, so I did not talk about it at all. I would avoid being involved with such newspaper publications, but I believe that it gave pleasure to Milana and she probably thought that it would give me pleasure too and that, in a way would help me to acquire certain rights vis-à-vis Einstein in people's eyes.[32]

Radmila Milentijević ends her book[33] by quoting Elisabeth Roboz Einstein, the second wife of Hans Albert, the first son of Albert and Mileva. She wrote how her husband was saddened by the thought of his mother. "The fact that her name had been omitted from Albert's publications, that her marriage had

[29] Milentijević, p. 142.
[30] Evan Harris Walker, "Did Einstein espouse his spouse's ideas?", and a rebuttal by John Stachel, *Physics Today*, Vol. 42, No. 2, p. 9 (1989).
[31] Trbuhović-Gjurić, p. 106.
[32] Popović, Mileva Marić's letter to Helena Savić, dated June 13, 1929, p. 158.
[33] Milentijević, p. 479.
[34] Elisabeth Roboz Einstein, *Hans Albert Einstein: Reminiscences of His Life and Our Life Together*, The University of Iowa, 1991, p. 3 (as cited by Milentijević, p. 479).

brutally come to an end and their son's disease, all this had a devastating effect on her life."[34]

Mileva Marić had her reasons to choose to remain silent. She, who was the first person to believe in Albert's potential, wrote to her friend Helena Savić in 1922, "Even my closest friends still feel a great deal of admiration for his scientific achievements and transfer that to the personal sphere. You alone understand me best when you were able to say: I no longer care for him ... "[35]

My opinion

Having read numerous books and articles dedicated to Mileva Marić, and in particular the biographies written by Desanka Trbuhović-Gjurić and Radmila Milentijević, the letters exchanged between Albert and Mileva at the beginning of their relationship, and the correspondence of Mileva Marić with her friend Helena Savić (edited by Milan Popović), and having consulted the very well-documented book by Dord Krstić, I have no doubts in my mind about their collaboration. This conclusion emerges only when several elements are combined, however, and does not rest on one unique, irrefutable piece of evidence. The best existing evidence is the reference made by Albert Einstein himself to their common research on the theory of relativity in his letter dated March 21, 1901, which I quoted earlier.

I think that the times and circumstances forced Mileva to put herself in the background, behind her husband. In addition, her complete love and trust in Albert were such that she supported him at all costs, delighted to participate in his success. By accepting that their common research should be published under his sole name, she made it possible for him to emerge and fully develop his talent. The price she had to pay was the sacrifice of her own career. If he had remained unemployed, Albert would not have married her. And once the stage had been set by mutual agreement, who would have been able to backtrack? Albert Einstein risked losing a lot: his professorial position, his fame, his name. The more he delayed rectifying the situation, the more he had to lose. His fame, acquired at her expense, probably ruined their beautiful initial team spirit. Eleven years after their separation, in 1925, after the relationship had changed and when it had become clear in Mileva's eyes that they no longer were one "single stone" as was the case at the beginning of their relationship, she tried to claim her dues. Albert's reaction was so brutal that she probably chose to keep silent forever. This would explain why Mileva refused to claim her share of the fame, even when her friend Milana Sefanović urged her to do so by speaking publicly about the matter in 1929.

---

[35] Popović, Mileva Marić's letter to Helena Savić, dated Zurich 1922, pp. 132–133.

The exact contributions of each of them will probably always remain a mystery. But everything seems to demonstrate that it was by working together that they were able to produce ideas of such creativity. All scientists who work within collaborations, in particle physics and elsewhere, know how beneficial such exchanges are. It does happen, of course, that a single person sometimes has a brilliant idea, but discussing such ideas with one's colleagues always pushes them a notch further. According to today's criteria, Mileva Marić would have been recognized as a coauthor of these theories. The historical context and the circumstances have decided otherwise.

Index